Geographical Research with 'Vulnerable Groups'

Drawing on varied expertise from specialisms across the sub-disciplines of social and cultural geography, this book seeks to interrogate what it is to do research with people widely considered to be vulnerable. Written from an emancipatory standpoint, this book addresses the ethical and practical challenges that face researchers working with marginalised people. With chapters exploring the authors' own experiences of working with a wide range of participants including homeless people, indigenous peoples, drug addicts, learning disabled children, and prisoners, the book draws on research undertaken by academics across the globe. *Geographical Research with 'Vulnerable Groups'* unpicks and interrogates each part of the research process, from obtaining ethics permission from review bodies, to recruitment and gatekeepers, through to dissemination of research findings. Throughout the discussion, authors foreground the relational identities of the actors in the research process, highlighting the ways in which institutional attempts to protect marginalised people from risk, perpetuate a perceived, and even material, vulnerability. This honest and empirically driven text will provide an illuminating insight for researchers embarking on research with marginalised people.

The chapters in this book were originally published as a special issue of *Social & Cultural Geography*.

Nadia von Benzon is a Lecturer in Human Geography at Lancaster University, UK. She teaches and researches social geography and qualitative methods, with a particular interest in issues of marginalisation and difference. Specific research interests include childhood, disability, health and wellness, and 'otherness'.

Lorraine van Blerk is Professor in Human Geography at the University of Dundee, UK. She has conducted research with marginalised young people including street-connected children and youth in Sub-Saharan Africa, young refugees in Uganda and Jordan as well as children living in poverty across a diverse range of contexts.

Geographical Research with 'Vulnerable Groups'

Re-examining Methodological and Ethical Process

Edited by
Nadia von Benzon and Lorraine van Blerk

Routledge
Taylor & Francis Group
LONDON AND NEW YORK

First published 2018
by Routledge
2 Park Square, Milton Park, Abingdon, Oxon, OX14 4RN, UK

and by Routledge
711 Third Avenue, New York, NY 10017, USA

Routledge is an imprint of the Taylor & Francis Group, an informa business

British Library Cataloguing in Publication Data
A catalogue record for this book is available from the British Library

ISBN13: 978-0-8153-9687-1

Typeset in MyriadPro
by diacriTech, chennai

Publisher's Note
The publisher accepts responsibility for any inconsistencies that may have arisen during the conversion of this book from journal articles to book chapters, namely the possible inclusion of journal terminology.

Contents

Citation Information

The chapters in this book were originally published in *Social & Cultural Geography*, volume 18, issue 7 (November–December 2017). When citing this material, please use the original page numbering for each article, as follows:

Chapter 1

Research relationships and responsibilities: 'Doing' research with 'vulnerable' participants: introduction to the special edition
Nadia von Benzon and Lorraine van Blerk
Social & Cultural Geography, volume 18, issue 7 (November–December 2017) pp. 895–905

Chapter 2

Relational vulnerability and the research process with former prisoners in Athens, Georgia (USA)
Matthew L. Mitchelson
Social & Cultural Geography, volume 18, issue 7 (November–December 2017) pp. 906–926

Chapter 3

Researching migration and enforcement in obscured places: practical, ethical and methodological challenges to fieldwork
Pauline Maillet, Alison Mountz and Keegan Williams
Social & Cultural Geography, volume 18, issue 7 (November–December 2017) pp. 927–950

Chapter 4

Civil society activists and vulnerability in South India: the relational politics of life history methods and development research
Matt Baillie Smith and Katy Jenkins
Social & Cultural Geography, volume 18, issue 7 (November–December 2017) pp. 951–970

Chapter 5

Being patient, being vulnerable: exploring experiences of general practice waiting rooms through elicited drawings
Kyle Eggleton, Robin Kearns and Pat Neuwelt
Social & Cultural Geography, volume 18, issue 7 (November–December 2017) pp. 971–993

Chapter 6

The ethnographic novel as activist mode of existence: translating the field with homeless people and beyond
Michele Lancione
Social & Cultural Geography, volume 18, issue 7 (November–December 2017) pp. 994–1015

Chapter 7

Residential ethnography, mixed loyalties, and religious power: ethical dilemmas in faith-based addiction treatment
Andrew P. J. W. Williams
Social & Cultural Geography, volume 18, issue 7 (November–December 2017) pp. 1016–1038

Chapter 8

Confessions of an inadequate researcher: space and supervision in research with learning disabled children
Nadia von Benzon
Social & Cultural Geography, volume 18, issue 7 (November–December 2017) pp. 1039–1058

Notes on Contributors

Matt Baillie Smith is a Professor of International Development at the Department of Social Sciences, Northumbria University, UK.

Kyle Eggleton is Senior Lecturer at the School of Population Health, The University of Auckland, New Zealand.

Katy Jenkins is an Associate Professor in International Development at the Department of Social Sciences, Northumbria University, UK.

Robin Kearns is a Professor at the School of Environment, The University of Auckland, New Zealand.

Michele Lancione is Senior Research Fellow at the School of Geography and Planning, Cardiff University, UK.

Pauline Maillet is Project Manager at VoxPublic, France.

Matthew L. Mitchelson is Assistant Chair of the Department of Geography and Anthropology and Associate Professor of Geography at the Kennesaw State University, USA.

Alison Mountz is a Professor at the Department of Geography and Environmental Studies, Wilfrid Laurier University, Canada.

Pat Neuwelt is an Honorary Academic at the School of Population Health, The University of Auckland, New Zealand.

Lorraine van Blerk is Professor in Human Geography at the University of Dundee, UK.

Nadia von Benzon is a Lecturer in Human Geography at Lancaster University, UK.

Andrew P. J. W. Williams is Lecturer in Human Geography at the School of Geography and Planning, Cardiff University, UK.

Keegan Williams was based at the Department of Geography and Environmental Studies, Wilfrid Laurier University, Canada.

Research relationships and responsibilities: 'Doing' research with 'vulnerable' participants: introduction to the special edition*

Nadia von Benzon and Lorraine van Blerk

ABSTRACT
For some decades now, human geography has sought to engage with the narratives of those deemed least powerful in society. This interest in the experiences of those who have traditionally been sidelined from political agendas has been part of move towards an increasingly critical approach to the social sciences, and ultimately as a way of using the academy as a tool for social justice. Researching the experiences of socially, politically and economically marginalised groups of people, is no longer a new endeavour. What is new, however, is the rapidity in the growth of interest of researchers seeking to engage with these populations, and the variety of the toolkit of methods, for research and analysis, available to support this engagement. Whilst this special edition seeks to draw out and explore a variety of novel approaches to research methods and practice, the particularly unique contribution presented by this publication is a candid exploration of the ethical challenges of engaging with populations identified as 'vulnerable' as expressed through the discussion of first-hand fieldwork experience.

Rapports et responsabilités de la recherche: 'faire' de la recherche avec des participants « vulnérables » : introduction à l'édition spéciale

RÉSUMÉ
Il y a maintenant quelques dizaines d'années que la géographie humaine tâche de s'intéresser aux discours de ceux qui sont considérés comme ayant le moins de pouvoir dans la société. Cet intérêt pour les expériences de ceux qui ont été traditionnellement écartés des projets politiques fait partie d'un mouvement vers une approche de plus en plus critique des sciences sociales et en fin de compte est une façon d'utiliser la recherche universitaire comme outil de justice sociale. Faire de la recherche sur les expériences de groupes de personnes marginalisées socialement, politiquement et économiquement n'est plus une démarche nouvelle. Ce qui est nouveau cependant, c'est la rapidité de l'augmentation de l'intérêt des chercheurs qui désirent entrer en contact avec ces populations et de la variété de la panoplie

*This Special Issue is based on the concept of a paper session at the Royal Geographical Society Annual Conference 2011 entitled: 'Doing Geography with Vulnerable People: Research, Methods and Practice'.

des méthodes, pour la recherche et l'analyse, disponible pour soutenir cet engagement. Cette édition spéciale cherche à mettre en lumière et explorer une variété d'approches nouvelles pour rechercher les méthodes et les pratiques et la contribution particulièrement exceptionnelle présentée dans cette publication est une exploration candide des défis éthiques de la mise en contact avec les populations identifiées comme « vulnérables » telles qu'elles sont articulées à travers la discussion de l'expérience à chaud sur le terrain.

Las relaciones y responsabilidades en la investigación: 'haciendo' investigación con participantes 'vulnerables': introducción a la Edición Especial

RESUMEN
Desde hace algunas décadas, la geografía humana ha tratado de comprometerse con las narrativas de aquellos que se consideran menos poderosos en la sociedad. Este interés en las experiencias de aquellos que tradicionalmente han estado al margen en las agendas políticas ha sido parte del movimiento hacia un enfoque cada vez más crítico a las ciencias sociales y, en última instancia, como una forma de usar la academia como un instrumento para la justicia social. La investigación de las experiencias de los grupos de gente social, política y económicamente marginados, ya no supone un nuevo esfuerzo. Lo que es nuevo, sin embargo, es la rapidez en el crecimiento del interés de los investigadores que buscan involucrarse con estas poblaciones, y la variedad del conjunto de herramientas de métodos, para la investigación y el análisis, disponibles para apoyar este compromiso. Si bien esta edición especial pretende extraer y explorar una variedad de enfoques novedosos de los métodos y la práctica de la investigación, la contribución particularmente singular que presenta esta publicación es una franca exploración de los desafíos éticos al involucrarse con poblaciones identificadas como 'vulnerables', como se expresa a través de la discusión de la experiencia de campo de primera mano.

Introduction

For some decades now, human geography has sought to engage with the narratives of those whose voices have been absent from research accounts and policy discourse. This interest in the experiences of those who have been sidelined from social, cultural, economic and political agendas has been part of move towards an increasingly inclusive approach to the social sciences and a means of using the academy as a tool for social justice. This approach is illustrated through geography texts since the 1990s, a new wave of 'critical geography' encompassing the burgeoning sub-disciplines of social and cultural geography. With roots in both a Marxist and a Feminist approach (Gibson-Graham, 1996; Massey, 1994; McDowell, 1999; Valentine, 2001), geographical engagement with the lived experiences of socioculturally, politically and economically marginalised groups of people, is no longer a new endeavour. What is new, however, is the rapidity in the growth of interest of researchers seeking to engage with these populations, and the variety of the toolkit of methods, for research and analysis, available to support this engagement. Whilst this special edition seeks to draw out and explore a range of novel approaches to research methods and practice, the particularly unique contribution presented by this publication is a candid exploration of the ethical challenges of engaging with populations identified as 'vulnerable' as expressed through the discussion of first-hand fieldwork experience.

The increased interest in research with people considered to be vulnerable mirrors a recent rapid development in interest and practice in social and cultural geography, and associated sub-disciplines such as children's geographies, the geography of education and geographies of disability (van Blerk & Kesby, 2009). What is clear from this body of work is that vulnerability is socially-constructed and dependent on the way in which power relations are created between mariginalised people, be they children, elderly, sick, disabled, or otherwise marginalised, and non-marginalised adults at a structural level across society (Pells, 2012; Tisdall & Punch, 2012). This special issue seeks to demonstrate the relational nature of vulnerability (see particularly Baillie Smith and Jenkins; and Eggleton, Kearns and Neuwelt, 2017). A relational approach demonstrates that vulnerability is context-dependent, with groups being more or less vulnerable to exploitation (in the widest sense of the word) based on the particular circumstances of an encounter. Therefore, research methodologies that are power-laden and unequal in their presentation and outcomes will position marginalised people as 'vulnerable'. At the heart of the methodological and ethical challenges to researching with groups who may be positioned to some degree as vulnerable, is the need to transform research from a 'top-down' researcher-led encounter to a 'bottom-up' participant-led encounter (Aldridge, 2014).

Qualitative approaches have been favoured for reflecting on the potentially unethical practices of some methods that may objectify marginalised groups, and geographers have increasingly sought to find new ways to engage in research relationships on a more equal footing (Hall & McGarrol, 2013; Moran, 2013; Parr & Stevenson, 2014; Williams, Cloke, May, & Goodwin, 2016). Participation has been increasingly valued as part of a qualitative approach where participant empowerment is viewed by researchers as a necessary, and beneficial, part of the research process (e.g. Askins & Pain, 2011; Cahill, 2007; Kindon, 2003). There is now a large literature that has employed qualitative and participatory research to highlight the experiences of diverse groups including those who may be considered vulnerable in a research encounter due to their age, status, education, impairments and other factors, (Kesby, Kindon, & Pain, 2005; Pain, 2004; Pain & Francis, 2003). Within this literature research that involves participants as decision-makers, and even researchers themselves, is widely understood to be the gold standard. Rigorous discussion and detailed explanation addresses all stages of conducting participatory research with marginalised people, from recruitment to research methods to analysis and dissemination.

Whilst there is no doubt that there exists a wealth of published discussion on the topic, this has not translated into straightforward practice in research with vulnerable groups throughout the academy. Despite a generalised willingness to do research alongside, rather than 'to', populations who have suffered discrimination or whose voices are less-well represented, human geographers are consistently faced with challenges to this endeavour. These challenges are explored in detail, in a variety of forms, within the papers contained in this special edition. Distinct from other special issues dealing with marginalised or vulnerable groups and research methodology, the papers here address a wide variety of different groups, in geographically diverse locations. However, the methodological and ethical challenges are unifying. These papers starkly indicate that the perception of inherent vulnerability, on the part of the people and the institutions that care for or contain them, the research institutions and the researchers themselves, puts strain on the research relationship, forcing researchers to compromise principles of 'bottom-up' working in order to work within limits imposed by institutions. Focused on a wide variety of international locations and contexts, this collection of papers shares an over-arching narrative: that research with people who experience vulnerability, or are perceived to be vulnerable, is complicated by issues concerning researcher–participant relations, and the responsibilities of the researcher, which are less fraught, and less contested, when researching our socio-economic peers.

This special edition acknowledges that conducting research with people who are considered vulnerable involves a more complex network of actors than is often the case in other research situations. Specifically, researchers intending to involve 'vulnerable' people as participants will need to engage with processes that may not allow for the specific tailored needs of the group being researched. For example, university ethics committees, who have power to permit or prevent research, are likely to put particular requirements on research practices (Baillie Smith and Jenkins, 2017; Dyer & Demeritt, 2009). Sometimes their procedures can be restrictive of the development of new ways of researching ethically with groups who do not fit the traditional model of research participants, or new ways of viewing participants as more than research subjects (Blake, 2007). Once permission is gained, most researchers will be confronted with other gatekeepers, usually in the form of host institutions, who may also put parameters on the practice of the research in order to ensure the protection of participants. Finally, researchers may be able to access the participants themselves, who may or may not be in a position to offer their own informed consent.

Through exploring the complexity of the research process we highlight how procedural issues of access and ethics in research with groups perceived to be vulnerable, elucidates further vulnerabilities. These vulnerabilities may, in turn interrupt or distort the way research is able to be carried out. Below we draw out three key foci that repeatedly emerge in this special issue and warrant further reflection from researchers: participant vulnerability; institutional vulnerability and researcher vulnerability. We discuss each of these in turn before exploring how the papers in this special issue bring these areas together in a call for more critical engagement in the research process and ethics. This critical engagement must be one that considers carefully the research relationships and responsibilities of all agential parties in the research process. As notions of vulnerability translate into risk, the perception of research with vulnerable people as a risky endeavour can significantly impact research.

Exploring 'vulnerability' in social and cultural geography research

Participant vulnerability

By and large minority, marginalised, and excluded populations are considered in social, political and economic terms as vulnerable. The notion of vulnerability reflects a socially constructed perception, and sometimes a reality, of a lack of social, political and economic capital held by such groups when compared to the societal norm. The groups explored in 2017: prisoners (Mitchelson), refugees and asylum seekers (Maillet, Mountz and Williams), indigenous peoples (Eggleton, Kearns and Neuwalt) disabled children (von Benzon), the homeless (Lancione),and drug addicts (Williams), share a common experience, where the societies in which they live perceive them to have reduced social and economic capital, which is seen to impede their capacity to act independently. Baillie Smith and Jenkins' participant group, of NGO activists in South India, presents an interesting contrast. In this case, the participant activists are not typically considered vulnerable people in their community context, but are relationally vulnerable in the context of government and intergovernmental development discourse and planning and their labour relations within this framework. Indeed, Baillie Smith and Jenkins' acknowledge that their participants included those who were wealthy and 'less vulnerable' by South Indian standards, but highly vulnerable when their labour situation was compared to that of a UK employee in a similar organisation. This underlines a key theme of this Special Issue, that vulnerability is not only material, but also relational.

Research already carried out with marginalised participant groups demonstrates that despite economic hardships or reduced social networks, new ways of developing social and economic capital emerge. For example, in discussing prisoners' experiences Moran (2013) highlights how the prison visiting room becomes a space where social networks from the outside can be re-engaged. Similarly, Johnsen, May, and Cloke (2008) discuss how auto-photography with homeless people facilitated an understanding of their everyday lives as similar to societal norms, in terms of activities carried out and social networks engaged with, albeit in different settings. Yet, the perceived lack of independence renders marginalised groups as reliant on others; whether they be carers, family members, adults, social workers, volunteers, health professionals or guards. This reliance on others, despite being a form of social capital in itself, creates the perception of vulnerability, and indeed materially generates socio-economic frailty, as individuals do not have control over their own assets, independent social networks, or relevant training and skills. This results in an institutional or policy perception of decreased social capital, and therefore decreased power, on the part of the 'vulnerable' adult or child.

This can be seen in everyday contexts in the UK, for example, through legislation such as that preventing children engaging in activities perceived as 'dangerous' that are permissible to adults, such as drinking or smoking, or the ability of medical professionals to incarcerate the mentally ill, or the 'rules' imposed on hospital wards that don't apply outside (see for example, Williams, 2017). This paternalism is well discussed and debated within disability studies, and to a lesser extent, geography of disability, where it is argued to be an unreasonable justification for the oppression of disabled people, and limitations to their rights. The argument is made eloquently in Wolpert's seminal paper 'The Dignity of Risk' where he argues that a paternalistic approach to disabled people, that prioritizes safeguarding over the rights of individuals to independent decision-making, is a limitation to personal freedom (Wolpert [1980]; see also work such as Tregaskis [2004]).

The same paternalistic desire to protect vulnerable people often extends to institutional policies that inhibit interaction between vulnerable clients and researchers (Carter, 2009). Mitchelson (2017) discusses the way in which historic abuses of prisoners through bio-medical and psychological experimentation have led to a widespread culture of institutional resistance to research participation from prisons. Mitchelson suggests that little distinction is made between the motivations of psycho-medical research and social science research, with institutions often on the offensive, presuming participants will be reduced to objectified subjects in academic discourse. Von Benzon (2017) in her research with learning disabled children, discusses similar frustrations illustrated through an exceedingly low response rate from families and from schools, apparently reticent to engage with any research process, prior even to having any knowledge of what the research might entail. Where institutions or individuals prevent researchers from accessing marginalised people, as a means of protecting the marginalised from harm, they prevent those in their care from contributing to a research agenda. In turn, this silences the voices of those who lack the independence to engage without 'permission' from gatekeepers. This lack of engagement limits the capacity of marginalised people to have either a direct or indirect influence on academic, and subsequently sociopolitical, understanding of their lived experience (Kramer-Roy, 2015).

This approach is completely at odds with a 'right-based' approach, or the concern of much of a new, critical approach in geography and the broader social sciences (see for example, the new sociology of childhood), that provides an argument for the necessity for marginalised people to be supported in decision-making about their own lives, rather than to have decisions made for them by others (Cope, 2008). Indeed, ideas such as 'Gillick Competence', applied in medicine, might provide a useful blueprint for determining whether people perceived as vulnerable should be given the opportunity to participate, and provide their own informed consent, in research projects. The concept of Gillick Competence, relates to the ability of children to consent to medical treatment independently of their parents once a child is able to show sufficient understanding and intelligence on the matter. This approach prioritises a relational understanding of intellectual maturity over an essentialist definition of age. Assessing individual capacity as a criteria for participation could be applied to many research contexts in which access to marginalised populations is frustrated by the actions of gatekeepers, due to the label of vulnerability placed on a potential participant, rather than any demonstrable inability (see for e.g. Parsons, Abbott, McKnight, & Davies, 2015).

Institutional vulnerability

Alongside a perception of marginalised people as 'vulnerable', care institutions are increasingly aware of their own vulnerability. We arguably live in a panoptic society, where the actions of protective and caring services, be they state owned, private companies, or within families, can be constantly monitored (Wrennall, 2010). Even private and domestic spaces are subject to surveillance through requirements to document, and the potential for whistle blowing. Such perpetual, or at least, potential, monitoring and interference, often leads those with a duty of care to marginalised people to be wary of increasing the number of 'visitors' and inviting potential whistle blowers inside. In turn, ethical challenges can be faced by researchers who do grapple with concerns of wrongdoing occurring within institutions in which they are working for fear of speaking up and undermining a research relationship and subsequently limiting the ability of individuals to participate (Maillet, Mountz and Williams, 2017; Williams, 2017).

Meanwhile Western society grows increasingly litigious, with service providers held accountable for wrongdoing (Cooper, 2005). In an increasingly risk averse society, institutions and families alike are bound to foreground a need to reduce the potential of harm to their charges, prioritising this over potential benefits of engagement in research. Gill (2007) particularly attributes restrictions on children's free play to a wide-spread social risk aversion. Whilst host institutions, acting as gatekeepers to marginalised people, may be the most obviously vulnerable institution in the research relationship, universities, as the institutions responsible for both the actions and the welfare of researchers, may also consider themselves vulnerable and act to minimise risk to themselves. Within this research relationship research institutions are at risk of litigation both from participants, or participating host institutions, and from the researchers themselves. Whilst participant vulnerability might manifest as an inability to act, perceived potential institutional vulnerability, underpinned by significant agency, will result in action to mitigate risk to both the potential research population, and perceived potential researcher vulnerability (Haggerty, 2004). A clear example of this in the Special Issue is Baillie Smith and Jenkins discussion of the use of consent forms in their research. These forms were insisted upon by the researchers' institution as a way of proving informed consent, but were protested against by participants who were reticent to sign a form that might be used to identify them.

Researcher vulnerability

A further vulnerability exists in the way in which researchers themselves are viewed as being at risk of harm from participants. Researchers may be viewed as particularly vulnerable when working with marginalised people due to a perceived increased risk of erratic or irrational behaviour on the part of the participant or indeed, more often, that the researcher will be exposed to emotional encounters that are upsetting and difficult to deal with. As Anderson and Smith (2001) highlight, emotions were, for a long time, ignored in the research process and only in the last decade or so have Geographers sought to explore this in more detail with Punch (2012, p. 86) noting that 'guilt, apprehension, fears and worries are legitimate, common and even useful experiences of fieldwork'. Bondi (2013) highlights the emotional benefits that can arise from engaging in research, not least for those who stories are more often ignored or hidden, but often researchers are less well equipped to deal with difficult and traumatic experiences raised through research (see for example, Mitchelson, 2017). For example, Briggs (2013) raises the emotional consequences of leaving the field when conducting visual ethnography with street drug users in London, emotions that are reflected in Williams' account (2017).

Further vulnerability may also be due to the fact that the participants are viewed as vulnerable, and unable to protect themselves, and therefore an 'easy target'. As such, a researcher could easily be perceived to have 'overstepped a mark' or acted inappropriately when behaving in a manner that would not have been labelled as such had the other person been considered a socio-economic peer, or superior. This is particularly a concern of ethics committees that seek to protect members of staff from harm through limiting their interaction with populations viewed as potentially harmful. This experience is discussed candidly in von Benzon's paper (2017), in which the author discusses the university ethics committee's requirement that two adults be present on research trips involving learning disabled child participants. Von Benzon reflects that this requirement on one hand, potentially provided another adult 'witness' when an incident did occur. On the other hand, she suggests, this incident might not have occurred, had the established, positive, one-to-one research relationship built up in the participant's home, been allowed to continue on the research vsisit.

This highlights the competing demands and expectations (Williams, 2017) that researchers are grappling with when in the field, which can further increase their emotional vulnerability, as they seek to develop relational research encounters with participants at the same time as they adhere to the institutional (university and host) requirements placed on them. As such, the researcher will enter the field with divergent responsibilities, or as Williams (2017) puts it, 'mixed loyalties'. By contrast, other researchers have embraced their own vulnerability in the field, drawing on the potential epistemological benefits of presenting their own vulnerability to participants and as such seeking to diminish relational vulnerability.

For Eggleton, Kearns and Neuwalt (2017), this is through a process of sharing the researcher's own drawing with the participants during the research process.

The final section of this paper seeks to explore some of the ways in which mixed loyalties may be competing through the lifecourse of research projects. It is hoped that by bringing these issues to the fore, the questions raised through the papers in the special edition may speak beyond the pages of this volume, as methods and ethics approaches are developed through research institutions and third sector organisations. Many of the papers in this special issue are highly personal accounts that candidly discuss the experiences of researchers in the field. At times these papers may be 'raw', drawing on the emotional encounters that researchers experienced and captured in their fieldwork diaries, which as Punch (2012) suggests are legitimate and revealing parts of the process. Voicing these struggles is something that is necessary for developing understanding, challenging current practice and shifting debates. This special issue intends to demonstrate the importance of honesty in methodological discussions, particularly regarding the specific challenges of doing research with people who are perceived as vulnerable (Horton, 2008). It is our hope, that through openness, and constructive self-criticism, not only will we as individual researchers learn from our mistakes, but others may be able to as well (Pyer, 2008).

Research relationships and responsibilities in fieldwork with vulnerable people

The papers in this special edition suggest that perceived participant vulnerability impacts on research projects from the earliest stages of conception and planning, through to the dissemination of research ideas (for example, Lancione, 2017). In particular these papers draw out the theme of research relationships, emphasising the complex mix of responsibilities navigated by geographers engaging with vulnerable people. In all of the papers in this special issue, authors engage with the notion of divided loyalties and competing responsibilities. Foremost in the researchers' interests remains the need to engage not only empathetically, but also positively, in the lives of the research participants. However, each paper alludes to the difficulties of meeting what researchers see as an ethical imperative, a responsibility or an obligation, whilst maintaining a positive and professional working relationship with or within a gatekeeper (institution) and/or their own university.

A number of the marginalised people who participated in the studies detailed in the special edition have little or no independence, due to institutionalisation (Williams, 2017; von Benzon, 2017) and incarceration (Maillet, Mountz and Williams, this issue) or heightened surveillance (Mitchelson, 2017). One key theme arising in the research is the issue of lack of privacy in research engagements when undertaking research with vulnerable people. This lack of privacy is largely a product of the institutionalisation of vulnerable people and/or the need to access these populations through institutional gatekeepers. This is particularly stark in Maillet et al.'s paper (2017) in which the authors discuss multiple instances of being forced to undertake interviews with incarcerated refugees and asylum seekers in the presence of guards. In this situation the capacity of the participant to take part in the research process was significantly limited through direct prohibitions on discussion topics enforced by guards, and/or through fear of reprisals if contributions were interpreted as complaints. Von Benzon (2017) similarly discusses the problem of outside interference in the research process, but in contrast finds this reduces the opportunities for participants to remove themselves from research interactions due to institutional pressure to conform to their alternate role as 'pupil'. Lack of privacy may also reflect a choice made by the researcher and the participants. In the case of Mitchelson (2017) a cafe was used as a meeting place that might be considered safe space for both parties. However, the contrast in these research experiences suggests that it is not the possibility of being overheard that is risky to the production of research data per se, but rather the potential of interference from those in positions of power over the participant.

Lancione's paper (2017) focuses on the broader duty of researchers engaging with vulnerable people to intervene in socially unjust situations through facilitating direct public engagement with the research outcomes. Lancione explains that for him, simply publishing his work in academic outlets wasn't sufficient. Society moves at a faster pace than academia, and Lancione's paper demonstrates the need to engage with non-traditional and non-academic means of dissemination, if research with marginalised people is to having any chance of having a positive impact on the lives of the vulnerable people who contributed. For Maillet, Mountz and Williams (2017) one means of facilitating broad dissemination will be through the use of a publicly accessible website, allowing access to research data and analysis to those outside academia. Disseminating research widely and effectively, for maximum impact as well as academic scholarship, is just one way in which researchers are able to attempt to ensure a positive outcome from research participation

Concluding remarks

This special edition seeks to bring together seven stories of doing geographical research alongside individuals, or groups, deemed 'vulnerable'. The papers explore the process of striving to be a responsible, or ethical, researcher, discussing candidly the challenges to this endeavour presented when the research participants are considered by others, and even themselves, to be vulnerable. Moreover, we draw out the vulnerability experienced by the researcher, and perceived by the sponsoring universities, that is distinctive to working with marginalised social groups.

As academics we recognise the burgeoning sub-disciplines within which many of the social and cultural geographers who've published in this special edition work. We value the creative and supportive communities within our subdisciplines, and the space these academic foci offer for the development of expert knowledge and networks. Whilst many of us may work within broad teams of human geographers within our institutions, we are often drawn to these specific communities for inter-institutional discourse such as conferences and workshops. This special issue seeks to draw out the similarities of research experience with very different participant groups, due to their shared identity as 'vulnerable'. As such, this special edition highlights the utility of broadening conversations beyond sub-disciplinary boundaries for knowledge sharing and potential discourse for change.

It is our hope that this special edition generates further reflection and honest discussion around the conflict that exists between the responsibility that institutions, whether academic or care-giving have to cause least harm, and the obligation felt by many social and cultural geographers to do more-than-least-harm. This is the recognition that 'doing' geography well may be a risky endeavour, and we need to do more as a discipline to convince our universities, and our potential host institutions that this is a risk worth taking. Where we take the easy way out and acquiesce to ethics committee requirements that we feel undermine our research aims, or allow gatekeepers to influence our method or practice, in fact we put much more at stake. In these circumstances we risk both our academic integrity, and finding a meaningful and ethical way of allowing marginalised people to 'do' geography alongside us.

Disclosure statement

No potential conflict of interest was reported by the authors.

References

Aldridge, J. (2014). Working with vulnerable groups in social research: Dilemmas by default and design. *Qualitative Research, 14*, 112–130.

Anderson, K., & Smith, S. (2001). Editorial: Emotional geographies. *Transactions of the Institute of British Geographers NS, 26*, 7–10.

Askins, K., & Pain, R. (2011). Contact zones: Participation, materiality, and the messiness of interaction. *Environment and Planning D: Society and Space, 29*, 803–821.

Baillie Smith, M., & Jenkins, K. (2017). Civil society activists and vulnerability in South India: the relational politics of life history methods and development research, *18*, doi: 10.1080/14649365.2016.1216157

Blake, M., (2007), Formality and friendship: Research ethics review and participatory action research *ACME: An International Journal for Critical Geographers, 6*, 411–421.

Bondi, L. (2013). Research and therapy generating meaning and feeling gaps. *Qualitative Inquiry, 19*, 9–19.

Briggs, D. (2013). Emotions, ethnography and crack cocaine users. *Emotion, Space and Society, 7*(1), 1–12.

Cahill, C. (2007). Including excluded perspectives in participatory action research. *Design Studies, 28*, 325–340.

Carter, B. (2009). Tick box for child? The ethical positioning of children as vulnerable, researcher as barbarians and reviewers as overly cautious *International Journal of Nursing Studies, 46*, 858–864.

Cooper, G. (2005). Disconnected Children. *ECOS, 26*, 26–31.

Cope, M. (2008). Becoming a scholar-advocate: Participatory research with children. *Antipode, 40*, 428–435.

Dyer, S., & Demeritt, D. (2009). Un-ethical review? Why it is wrong to apply the medical model of research governance to human geography. *Progress in Human Geography, 33*, 46–64.

Eggleton, K., Kearns, R., & Neuwelt, P. (2017). Being patient, being vulnerable: exploring experiences of general practice waiting rooms through elicited drawings, *18*, doi: 10.1080/14649365.2016.1228114

Gibson-Graham, J. K. (1996). *The end of capitalism (as we knew it): A feminist critique of political economy*. Oxford: Blackwell.

Gill, T. (2007). *No fear: Growing up in a risk averse society*. London: Calouste-Gulbenkien Foundation.

Haggerty, K. (2004). Ethics creep: Governing social science research in the name of ethics. *Qualitative Sociology, 27*, 391–414.

Hall, E., & McGarrol, S. (2013). Progressive localism for an ethics of care: Local area co-ordination with people with learning disabilities. *Social and Cultural Geography, 14*, 689–709.

Horton, J. (2008). A 'sense of failure'? Everydayness and research ethics. *Children's Geographies, 6*, 363–383.

Johnsen, S., May, J., & Cloke, P. (2008). Imag(in)ing 'homeless places': Using auto-photography to (re) examine the geographies of homelessness. *Area, 40*, 194–207.

Kesby, M., Kindon, S., & Pain, R., (2005), Participatory approaches and diagramming techniques, In: R. Flowerdew & D. Martin (Eds.), *Methods in human geography: A guide for students doing a research project*. Essex: Pearson.

Kindon, S. (2003). Participatory video in geographic research: A feminist practice of looking? *Area, 35*, 142–153.

Kramer-Roy, D. (2015). Using participatory and creative methods to facilitate emancipatory research with people facing multiple disadvantage: A role for health and care professionals. *Disability and Society, 30*, 1207–1224.

Lancione, M. (2017). The ethnographic novel as activist mode of existence: translating the field with homeless people and beyond, *18*, doi: 10.1080/14649365.2016.1231336
Maillet, P., Mountz, A., & Williams, K. (2017). Researching migration and enforcement in obscured places: practical, ethical and methodological challenges to fieldwork, *18*, doi: 10.1080/14649365.2016.1197963
Massey, D. (1994). *Space, place and gender*. Minneapolis: University of Minnesota Press.
McDowell, L. (1999). *Gender, identity and place: Understanding feminist geographies*. Cambridge: Polity.
Mitchelson, M.L. (2017). Relational vulnerability and the research process with former prisoners in Athens, Georgia (USA), *18*, doi: 10.1080/14649365.2016.1171388
Moran, D. (2013). Between outside and inside? Prison visiting rooms as liminal carceral spaces *GeoJournal, 78*, 339–351.
Pain, R. (2004). Social geography: Participatory research. *Progress in Human Geography, 28*, 652–663.
Pain, R., & Francis, P. (2003). Reflections on participatory research. *Area, 35*, 46–54.
Parr, H., & Stevenson, O. (2014). No news today: Talk of witnessing with families of missing people. *Cultural Geographies, 22*, 297–315.
Parsons, S., Abbott, C., McKnight, L., & Davies, C. (2015). High risk yet invisible: Conflicting narratives on social research involving children and young people, and the role of research ethics committees. *British Educational Research Journal, 41*, 709–729.
Pells, K. (2012). 'Rights are everything we don't have': Clashing conceptions of vulnerability and agency in the daily lives of Rwandan children and youth. *Children's Geographies, 10*, 427–440.
Punch, S. (2012). Hidden struggles of fieldwork: Exploring the role and use of field diaries. *Emotion, Space and Society, 5*, 86–93.
Pyer, M. (2008). Unintended consequences? Exploring the un(fore)seen effects and outcomes of research. *Children's Geographies, 6*, 213–217.
Tisdall, K., & Punch, S. (2012). Not so 'new'? Looking critically at childhood studies *Children's Geographies, 10*, 249–264.
Tregaskis, C. (2004). Applying the social model in practice: Some lessons from countryside recreation. *Disability and Society, 19*, 601–611.
Valentine, G. (2001). *Social geographies: Society and space*. Harlow: Longman.
van Blerk, L., & Kesby, M. (2009). *Doing children's geographies: Methodological issues in research with young people*. London: Routledge.
von Benzon, N. (2017). Confessions of an inadequate researcher: space and supervision in research with learning disabled children, *18*, doi: 10.1080/14649365.2016.1257148
Williams, A., Cloke, P., May, J., & Goodwin, M. (2016). Contested space: The contradictory political dynamics of food banking in the UK. *Environment and Planning A, 48*, 2291–2316.
Williams, A. P. J. W. (2017). Residential ethnography, mixed loyalties, and religious power: ethical dilemmas in faith-based addiction treatment, *18*, doi: 10.1080/14649365.2016.1247194
Wolpert, J. (1980). The dignity of risk. *Transactions of the Institute of British Geographers, 5*, 391–401.
Wrennall, L. (2010). Surveillance and child protection: De-mystifying the Trojan horse. *Surveillance and Society, 7*, 304–324.

Relational vulnerability and the research process with former prisoners in Athens, Georgia (USA)

Matthew L. Mitchelson

ABSTRACT

A host of power relations are structurally imposed upon the imprisoned and these produce vulnerabilities that require considerable energy and attention. This paper argues, however, that 'vulnerable populations' is a politically reactive legal category in the context of research with prisoners. In turn, this form of categorical vulnerability is often misaligned with 'actually existing vulnerability' or 'relational vulnerability' in the context of prison life. Substantive ethical, methodological and practical problems result from this misalignment, as I illustrate through the case of my research with former prisoners in Athens, Georgia. I seek to prove two claims in support of my central argument. First, actually existing vulnerability is fundamentally relational and it involves the researcher, the researched, collaborators, and (often distant) people and social structures that were not necessarily directly involved in the research process. Second, vulnerability is a spatially and temporally contingent process. The essay reviews the peculiar history by which prisoners came to be categorized as a 'vulnerable population' in the United States, in addition to recounting my first-hand experiences during the research process.

Vulnérabilité relationnelle et procédé de recherche auprès d'anciens prisonniers d'Athens en Géorgie (USA)

RÉSUMÉ

Une grande quantité de rapports de force sont imposés structurellement à ceux qui sont emprisonnés et ces rapports produisent des vulnérabilités qui demandent une énergie et une attention considérables. Cet article soumet, toutefois, que « les populations vulnérables » forment une catégorie légale politiquement réactive dans le contexte de la recherche auprès des prisonniers. A son tour, souvent, cette forme de catégorie de vulnérabilité ne correspond pas à « la vulnérabilité réelle qui existe » ou à la « vulnérabilité relationnelle » dans le contexte de la vie en prison. Des problèmes substantiels éthiques, méthodologiques et pratiques résultent de ce décalage, comme je le démontre à travers le cas de ma recherche auprès d'anciens prisonniers d'Athens, en Géorgie. Je cherche à prouver deux affirmations pour appuyer mon argument central. Tout d'abord, la vulnérabilité qui existe réellement est fondamentalement relationnelle et elle inclue le chercheur, ceux sur qui la recherche est faite, les collaborateurs et les structures sociales et humaines (souvent distantes) qui n'étaient pas forcément directement impliquées dans le processus de recherche. Deuxièmement, la vulnérabilité est un processus variable spatialement et temporellement. L'article fait un compte rendu de l'étrange développement qui a mené à la catégorisation des prisonniers comme « population vulnérable » aux Etats-Unis, et fait aussi le récit de mes expériences personnelles pendant le processus de recherche.

Vulnerabilidad relacional y proceso de investigación con antiguos presos en Athens, Georgia (EE.UU.)

RESUMEN

Una gran cantidad de relaciones de poder se imponen estructuralmente a los presos y estas producen vulnerabilidades que requieren energía y atención considerables. Este documento sostiene, sin embargo, que las "poblaciones vulnerables" es una categoría jurídica políticamente reactiva en el contexto de la investigación con los presos. A su vez, esta forma de vulnerabilidad categórica es a menudo mal alineada con "vulnerabilidad real" o "vulnerabilidad relacional" en el contexto de la vida en prisión. Problemas éticos, metodológicos y prácticos sustanciales son el resultado de esta falta de alineación, como se ilustra a través del caso de la investigación con ex- prisioneros en Athens, Georgia. Se busca demostrar dos cuestiones para apoyar el argumento central. En primer lugar, la vulnerabilidad realmente existente es fundamentalmente relacional e implica al investigador, a los investigados, a los colaboradores, y (a menudo distantes) personas y estructuras sociales que no necesariamente estaban directamente involucradas en el proceso de investigación. En segundo lugar, la vulnerabilidad es un proceso espacial y temporalmente contingente. El ensayo revisa la historia particular en la que los prisioneros llegaron a ser categorizados como una "población vulnerable" en los Estados Unidos, además de relatar experiencias de primera mano durante el proceso de investigación.

Introduction

Vulnerability is a fraught concept in the context of research on imprisonment. Legal studies (e.g. Hoffman, 2000; Kalmbach & Lyons, 2003; McCarthy, 1989) and commentaries in clinical research fields (e.g. Lerner, 2007; Levine et al., 2004; McDermott, 2013) consistently argue that categorical vulnerability and its application are generally misaligned with societal needs. These misalignments stem in part from the term's widespread adoption and use without a formal definition. Levine et al. (2004, p. 45) note, for example, that:

> The U.S.-Code of Federal Regulations does not define vulnerability but provides special protections for 'particularly vulnerable populations,' specifically pregnant women, human fetuses, and neonates; prisoners; and children. Although used freely in the Institutional Review Board (IRB) Guidebook offered by the Office of Human Research Protections, the term vulnerability is not defined in its extensive glossary.

Of the particularly vulnerable populations alluded to here, only prisoners are defined by their location. Important geographic questions are as yet overlooked in the research ethics and carceral geography literatures, and this essay seeks to address two. First, how and why did categorical 'vulnerability' get here (i.e. placed within the prison)? Second, what are the implications (i.e. ethical, methodological and practical) of vulnerability's placement within the prison for contemporary human-geographic research?

The answer to the first question is seemingly straightforward, because prisons are violent, coercive institutions. People in prisons are unambiguously vulnerable. For example, prisoners have been exposed to various harms – including death – as part of twentieth century medical research in the United States that engaged participants with: cholera, pellagra, gonorrhoea, gas gangrene, dengue fever, malaria, herpes, pappataci fever and sleeping sickness; injections of blood from beef cattle, testicular transplants and testicular radiation, plutonium injections, and radiation exposure (Hornblum, 1997; Kalmbach & Lyons, 2003; Lerner, 2007). Two cohorts of political reactions – what I term the post-Nuremberg and post-Belmont eras of research – are central to understanding categorical vulnerability's evolution and placement within the prison. Both eras consist of political reactions to highly visible ethical violations involving clinical/medical research – research that is often methodologically incommensurable with research in the humanities and social sciences (cf. Burr & Reynolds, 2010). Yet, these political reactions have firmly entrenched 'vulnerability' in universalizing, categorical legal terms in the context of imprisonment for all researchers who are (directly or indirectly) subject to federal regulations and local IRB oversight.

When 'vulnerability' is reproduced and accepted in this way, the regulation of this legal category (and others) by IRBs often structurally and problematically positions human beings in very specific, ostensibly manageable ways (Bradley, 2007; Martin & Inwood, 2012; Schneider, 2015). Prisoners are discursively positioned as universally 'at risk' to research harms. In turn, their agency is discursively eliminated and they are often practically excluded from research that may potentially be of direct benefit to them (Hoffman, 2000). Researchers are discursively positioned as universally 'risky', powerful wielders of those potential harm(s). Discursively, most historical analyses of research ethics abuse(s) clearly 'treat the research subjects as victims and the researchers as oppressors' (Comfort, 2009 p. 191). And local IRBs are appointed as the adjudicators of a given research design's acceptability, effectively serving as their institution's research 'risk manager'. Each of these subject positions has been hotly contested in recent years, as part of larger debates about IRBs and research oversight. For example, Schneider has argued, 'prisoners are vulnerable; prisons are brutal; researchers have abused prisoners. But it does not follow that IRBs know what is best for prisoners' (2015, p. 133).

In this essay, I argue that the politically reactive, legal category of 'vulnerable populations' is highly problematic and merits investigation in the context of research on imprisonment. Specifically, this categorical use of 'vulnerability' is routinely out of alignment with 'actually existing vulnerability' (cf. Brenner & Theodore, 2002), or, what can be usefully labelled and understood through the lens of 'relational vulnerability' (Few & Rosen, 2005; Few-Demo & Arditti, 2014; Huijsmans, 2012; Kabeer, 2014; Kabeer, Khawar, & Sayeed, 2010; Taylor, 2013). This misalignment often creates substantive ethical, methodological, and practical problems and complications that are as yet underexplored in the research ethics and methods literature regarding the geographies of imprisonment. A case can be made that relational vulnerability is often implicitly present (and at work) in the prison ethnography literature (Jewkes, 2011;

Ugelvik, 2014a; and see Qualitative Inquiry (volume 20) in particular). It is my hope that making relational vulnerability's implications explicit can complement that work, although the research experience I report on below is not a prison ethnography per se. Like one prison ethnographer, I believe that 'the researcher's lived experiences, including her or his situated emotions and feelings, are the central methodological tools available' (Ugelvik, 2014b, p. 479) in the work that I have done. Like another, I aspire to conduct human-geographic research with a 'concept of personhood' that moves beyond categorical thought (e.g. I aspire to partner with co-producers of knowledge; not wield power over vulnerable subjects) and I hope to 'humanize our understanding of the prison and the prisoner in ways that are of deep significance not just to penalogical scholarship, but also to late modern fractured communities' (Liebling, 2014, p. 484).

I seek to make two points in support of my summative argument (i.e. that categorical vulnerability is problematic, however well-intentioned it may be), and to explore these themes through my work with *former* prisoners (see below). First, actually existing vulnerability is 'fundamentally relational' (cf. Watts & Bohle, 1993 p. 54). While researchers and other participants do not engage in the research process or 'the field' (Katz, 1994) on equal footing, their engagement is shared nonetheless. Research is a relational process, however asymmetrical its relationship(s) may be. In other words, *vulnerability neither started nor ended with the former prisoners with whom I worked*. So, the list of vulnerable participants in a given research process, when considered through the lens of relational vulnerability, must include the researcher too, as well as their collaborators and, at times, other people and social structures that may not be directly involved in the research process. Second, by virtue of its relationality, actually existing vulnerability is a spatially and temporally contingent process. Treating all prisoners as a 'vulnerable population' applies a universal social category to people in a space that is treated as fixed and static in the case of the prison, whereas imprisonment entails a dynamic human geography (Baer & Ravneberg, 2008; Gilmore, 2007; Martin & Mitchelson, 2009). In other words, *my participants' vulnerabilities neither started nor ended at their prison's walls*. For example, I found that various levels of vulnerability characterized the lives of my participants before their incarceration(s), during their time as a prisoner (i.e. as a member of a technically 'vulnerable population'), and that it unambiguously characterizes their lives during 'reentry' (Petersilia, 2001; Travis, 2001).

The rest of this essay is organized in three main parts. In the first part, I develop the concept of relational vulnerability. I briefly treat the historical roots of categorical vulnerability in the context of US imprisonment research, so as to understand the protectionist framework by which contemporary IRBs approach imprisonment research. I then contrast categorical vulnerability with an extended discussion of relational vulnerability and its ethical, methodological, and practical potential(s). In the second part, I present a case study and discussion of my work with former prisoners in Athens, Georgia. In so doing, I hope to illustrate the utility of adopting a relational approach to vulnerability. I also hope to account for the multiple, overlooked vulnerabilities associated with my research outside of prisons, and to help fill the practical 'gap in the methods literature' in imprisonment research (Sloan & Drake, 2013, p. 24). In the third part of this essay, I briefly summarize my argument and connect it to broader debates about vulnerability and research ethics while signalling areas for future research. I turn now to a brief history of categorical vulnerability's birth at the federal level in the United States, its placement within the prison, and its regulation by IRBs before expanding on what I mean by relational vulnerability.

Relational vulnerability and the prison: placing and producing prisoners as a 'vulnerable population'

The contemporary IRB framework for mitigating prison-related vulnerabilities in humanities and social science research is problematic for several reasons. First, this framework is premised on fixed, essentialized social categories that effectively dehumanize and depersonalize what are always human research processes. Second, this framework is premised on a fixed, essentialized understanding of prison space. The roots of this contemporary framework are historically grounded in what I term the post-Nuremberg and post-Belmont eras of imprisonment research. As a brief historical treatment of this era will illustrate, categorical vulnerability results from political reactions to research ethics abuses in clinical/medical contexts that are often incommensurable with the humanities and social sciences, and are often premised on peculiar and problematic geographic imaginaries.

Post-Nuremberg research

Nazi experimentations on prisoners in concentration camps are likely the most notorious known abuses of prisoners 'in the name of science', although abusive medical experimentations on prisoners have been documented for millennia (Kalmbach & Lyons, 2003, p. 672). The Nuremberg Medical Trial's relationship to research and imprisonment appears to be straightforward. The Trial (1946–1947) established the Nuremberg Code, which establishes uninformed experimentation without consent – with prisoners, specifically – as a 'crime against humanity' (quoted in McCarthy, 1989, p. 58). Further, the Code's opening consent clause mandates participants' 'free power of choice, without intervention of any element of force ... constraint or coercion' (quoted in Harkness, 1996, p. 1672). Yet, prison research expanded dramatically in the wake of the Nuremberg Medical Trial.

One could reasonably expect a decline in prison experimentations in the post-Nuremberg United States for at least three reasons. First, because imprisonment is an unambiguously 'coercive intervention' research on prisoners would often be considered a violation of the Nuremberg Code. Second, the horrors of Nazi experimentation on prisoners in the camps were widely publicized, as were the executions of four medical doctors and three Nazi officials charged during the Trial. Third, prewar prisoner experimentation had held a liminal, marginal status. Hornblum (1997, p. 1438) has argued that prewar prison research in the United States was viewed by 'the orthodox medical community' as 'the preserve of unsophisticated medical eccentrics investigating offbeat scientific theories'. On the other hand, wartime research with prisoners had proven too valuable, and selectively profitable, to be abandoned (Hornblum, 1997); it had established a 'cottage industry' (Lerner, 2007, p. 1806; McCarthy, 1989). Reverby (2013, pp. 101, 331) has argued that 'the horrors of the experiments under the Nazis became thought of as Nazi science, not science' and 'American research and Nazi murder' were rhetorically 'set up as opposites'. So, the emergence of 'dual codes' (Hornblum, 1997,p. 1439) – one for the world under the Nuremberg Code, and one for the United States – resulted indirectly from the Trial. These dual codes served the political function of positioning 'exceptional' American experimentation as incommensurable to Nazi offenses, and the economic function of effectively insulating medical interest from the Nuremberg Code's purview on ethical grounds.

The production of American exceptionalism and these dual codes required a considerable political effort, and Harkness (1996) has chronicled the ethically dubious sequence of events by which they were produced. Defense attorneys unexpectedly compared Nazi experimentation to wartime malaria research in an Illinois state prison (cf. Comfort, 2009) during the trial. Andrew C. Ivy, attended the Trial as an official American Medical Association (AMA) consultant 'for expert advice on matters of medical science and medical ethics' (Harkness, 1996, p. 1673), and he contacted Illinois Governor Dwight H. Green about convening an advisory committee. The committee consisted of three physicians (including Ivy), a rabbi, a priest and Oscar Mayer (a meatpacking company president), and it met in November and December of 1947. However, Ivy testified as a rebuttal witness in the Nuremberg Medical Trial in June of 1947. Five months prior to the committee's first meeting, Ivy 'implied that the Green Committee had carefully considered and approved the [Illinois prisoner] research; he never volunteered that the prestigious Green Committee had never met' (Harkness, 1996, p. 1674). Throughout the proceedings, Ivy was adamant that US-based prison experimentation was morally or ethically incomparable to Nazi experimentation in the camps.

Ivy's position was publicly sanctioned shortly thereafter. The *Journal of the American Medical Association* (JAMA) published an oft-cited 'endorsement' (McDermott, 2013) of Ivy's position, and the Green Committee's 'special article' lauded malaria research with Illinois prisoner as 'ideal' (Committee Appointed by Governer Dwight H. Green of Illinois, 1948, p. 457). The Green Committee's account has drawn considerable criticism in subsequent years. Comfort, for example, argues that:

> Indeed, counts 2 and 3 in the allegations of war crimes against the Nazi doctor, Karl Brandt [Hitler's personal physician], included a description that differs from the Stateville project only in the magnitude of the resulting deaths ... many of the Stateville experiments violated the Nuremberg Code ... Any candid interpretation of the Stateville malaria experiments must conclude that in some ways they were clearly unethical by the international standards of the day (2009, p. 200).

At the time of the Green Committee's report, the *JAMA* editor was Morris Fishbein, a Chicago cardiologist. He was also a member of the Green Committee (Harkness, 1996, p. 1675). Regardless of the Green Committee's ethical legitimacy, its practical implications clearly sanctioned research with prisoners in the United States. More than 90% of investigational drug tests in the United States were conducted on prisoners by the early 1970s (Kalmbach & Lyons, 2003, p. 673). By 1980, however, roughly 15% of investigational drug tests in the United States were conducted on prisoners (Kalmbach & Lyons, 2003, p. 674). This drastic shift in investigational drug testing in the prison signals what amounts to a research revolution in the United States, or the transition into what I term the post-Belmont era.

Post-Belmont research

If the Green Committee report's publication opened a floodgate of prison research in the post-Nuremberg era, it was effectively closed by the Belmont Report and other responses to twentieth century research scandals, many of which involved prisoners. The era of American exceptionalism ended abruptly. Medical research in prisons was 'effectively banned' by the late 1970s (McDermott, 2013, p. 10), and this trend has largely remained in place. The National Academy of Sciences' Committee on Ethical Considerations for Revisions to DHHS Regulations for Protection of Prisoners Involved in Research (Institute of Medicine of the National Academies, 2006, p. 116) has argued that:

> The perception that all forms of research involving prisoners are equally ethically problematic, or subject to blanket prohibition or to conditions so onerous that the research is not worth doing, may be responsible for the dwindling of prisoner research participation to the point that justice concerns have been expressed about the exclusion of prisoners from clinical trials.

US-based imprisonment research in the humanities and social sciences – which had surged alongside clinical research in the post-Nuremberg era – was comparably curtailed in the post-Belmont era. Reiter argues that, 'Twenty-first century prison research is scarce relative to (a) twentieth century prison research and (b) to the scale of twenty-first century US imprisonment' (Reiter, 2014, p. 417). And, as Wacquant has put it, 'the ethnography of the prison in the United States is not merely an endangered species but a virtually extinct one' (Wacquant, 2002, p. 385).

Two peculiar historical socio-spatial imaginaries were particularly important to the creation of Belmont: Nuremberg and Tuskegee. The first paragraph of the Belmont Report invokes the Nuremberg Code, and it goes on to establish the 'three fundamental ethical principles required of all human experimentation' (Kalmbach & Lyons, 2003, p. 677). The US Public Health Service (PHS) conducted the infamous 'Tuskegee Study of Untreated Syphilis in the Negro Male' from 1932 to 1972, during which 399 African-American men with late-stage syphilis (and a 201-man control population) were neither ethically informed of, or treated for, their infections (Heller, 1972). Public outcries resulted, as did sweeping federal regulations. Reverby argues that "Tuskegee' became 'America's Nuremberg' and seemingly brought the issues raised by the Nazi experiments 'home' for the first time' (2013, p. 337). For her:

> ... the [Tuskegee] Study entered a historical fog. Alabama became a prison, the men were concentration camp victims, and the PHS physicians morphed into Nazis. The underlying focus on race that suffused the Study, and its location deep within the Alabama Black Belt, often became a way to distance researchers from the PHS physicians.

Peculiar geographic imaginations become visible within this 'historical fog', whereby violations of 'fundamental ethical principles' can be clearly named (as happening 'in' Nuremberg and Tuskegee) but only ambiguously placed in space and time. In other words, very specific abuses of research ethics morphed into highly debated and contested historical anecdotes that were performed by distant offenders in either Nazi Germany or the Deep South. The Tuskegee Study did not involve prisoners. Yet, it would be consistently invoked in establishing the subsequent federal protections afforded to all prisoners as research subjects. In the wake of Tuskegee, the political reaction to threats represented by the discourse of Nuremberg and Tuskegee assumed an unambiguously 'protectionist perspective' (Institute of Medicine of the National Academies, 2006, p. 116). Research subjects were to be protected, and some subjects were more categorically vulnerable than others.

Ultimately, specificity and ambiguity are both at work in the use of categorical vulnerability that characterizes post-Belmont research and this creates a host of problems. Prisons, specifically, are implicitly identified as geographically vulnerable locations. Prisoners are specifically defined as categorically vulnerable populations due to their location within a prison. Yet, vulnerability remains formally undefined – so the unambiguously *human* link between imprisonment and research remains undefined. I have argued that many of categorical vulnerability's ethical, methodological and practical inadequacies result from this *ontological commitment to social and locationalcategorization* in the context of prison research. What are the potential implications of re-placing and reimagining vulnerability in the context of imprisonment, so as to more fully account for prison life?

Relational vulnerability and the prison

Judith Butler (2004) has argued for a humanizing retheorization of vulnerability in social terms. She advocates a reimagining of vulnerability in primal terms: a 'particular vulnerability, a vulnerability to the other that is part of bodily life, a vulnerability to a sudden address from elsewhere that we cannot preempt' (p. 29) and 'a common human vulnerability, one that emerges with life itself' (p. 31). Butler is careful to distinguish this 'common' vulnerability from a universalizing, categorical vulnerability, in part because 'vulnerability takes on another meaning the moment it is recognized, and recognition wields the power to reconstitute vulnerability' (p. 42). Or, if there is a 'thing' we call vulnerability, it is also a process. Prisoners are undoubtedly vulnerable at many moments in time. Their vulnerability can be understood as a 'thing' in this sense, and this 'thing' does structure their lives. In this sense, prisoners are categorically vulnerable. However, applying Butler's concept of this 'common' vulnerability to the context of imprisonment research, I would argue that prisoners are not only categorically vulnerable; their vulnerability is not only a 'thing' possessed by all prisoners. Instead, this form of *vulnerability emerges as the socio-spatial process of imprisonment unfolds* (Martin & Mitchelson, 2009). The always human geography of imprisonment is more complex than it is often represented to be (Martin & Mitchelson, 2009; Mitchelson, 2014).

Relational vulnerability, as I intend it, thus involves a dialectical epistemology of the social and spatial relationships that produce vulnerability, within and beyond the prison (Jones & Popke, 2010; Merrifield, 1993; Mitchelson, 2014; Soja, 1980). 'Relational vulnerability' has been previously used in several disparate research contexts to combat the oversimplifications and essentialism associated with non-dialectical epistemologies and analysis. Huijsmans (2012, p. 30), for example, has sought 'to transcend some of the artificial boundaries' that commonly categorize the agency and vulnerability of migrants in the field of youth studies. Taylor (2013, p. 318) has considered the dialectical relationship between vulnerability (for some) and security (for others) in the context of climate change from a political ecology perspective. Scholars of intimate partner violence (Few & Rosen, 2005; Few-Demo & Arditti, 2014) have used the term to consider intimate relationships alongside broader 'situational' contexts (e.g. life circumstances and life stage stressors). Kabeer (2014; Kabeer et al., 2010), working in the field of gender and development, has used the concept to address the endemic vulnerabilities associated with violence against women that are too often invisible or absent from discourses of human rights and human development. Although these studies have several implicit commonalities, there is no shared definition of the term between them, and the concept has yet to explicitly appear in the research ethics or carceral geography literature. What they share, I would argue, is an epistemological commitment to socio-spatial dialectics. A 'thing's' context always matters as a process unfolds. And the social and spatial are understood as interdependent, or, co-constitutive of a process that produces vulnerability. I am not arguing against claims that prisoners are categorically vulnerable as a result of their location. Prisons are unambiguously violent, coercive environments. Instead, I am arguing that prisoners are not *only* categorically vulnerable as a result of their location.

There are more than two million people confined in the United States on a given day, and each individual human being within a cage is vulnerable. The prison itself is most often a rigidly defined place, an architectural fact lined with razor wire. And, to be certain, any number of horrors has taken place within these places. This is true in part because the dynamic, always human geography of imprisonment is seldom confined to the prison's walls. Indeed, most carceral-geographic work approaches the prison through a lens of geographic relationality (Mitchelson, 2014) for which the 'unit of analysis' is seldom limited to the prison itself. Instead, geographers have theorized, identified and analysed the socio-spatial process of imprisonment through: 'transcarceral spaces', which extend 'beyond the physical walls of the prison' (Moran, 2013, p. 13); 'heterotopias', which, after Foucault, account for the 'inside/outside distinctions and indistinctions' of prison space (Baer & Ravneberg, 2008, p. 214); and 'abandoned regions', which constitute 'a single – though spatially discontinuous' arrangement of places from which prisoners come, and to which they go (Gilmore, 2008, p. 31). In all of this work, there is an implicit emphasis on 'prison life', which spatially 'opens' the prison to account for 'the everyday workings of incarceration as they are necessarily lived and experienced, as a form of human life and struggle, inside and outside prisons' (McKittrick, 2012, p. 956). Thus, as prisoners are not only categorically vulnerable as a result of their location, they are not the only ones who encounter (and produce) vulnerability in the context of imprisonment.

Prison ethnographers consistently grapple with power dynamics (and, in turn, different forms of vulnerability) in implicitly dialectical terms. As Liebling puts it, 'power flows overtly, and we become infused with as well as subject to it' (2014, p. 482). Prisons are generally restricted environments, and 'access' to its restricted spaces is often characterized as 'a process, not a moment' (Goodman quoted in Reiter (2014, p. 421)), and this process entails a great deal of strategic and/or political effort in which the researcher is ultimately beholden to institutional gatekeeping (Bosworth, Hoyle, & Dempsey, 2011). This requires that prison researchers often mitigate challenging power dynamics across the space of the prison itself, in relation to its prisoners, and in relation to their imprisoners (Crewe, 2014; Crewe, Warr, Bennett, & Smith, 2014; Drake & Harvey, 2014; Reiter, 2014; Sloan & Drake, 2013). In turn, this demands close attention to 'agency, identity management, and socialization and survival strategies' within a particularly difficult setting (Jewkes, 2011, 2014, p. 387). During his fieldwork in Norway's largest prison, Ugelvik argued that 'the field was basically a minefield' (Ugelvik, 2014a, p. 35, 2014b).

As Rowe (2014, p. 411) explains, 'themes of identity and power run together hand-in-hand – always, but especially in prisons, where no-one ever just happens to be'. Unlike many others involved in a carceral setting, researchers voluntarily enter (and leave) the prison. In turn, prison ethnographers consistently acknowledge that 'the very limited experience of the researcher *cannot compare* to the all-encompassing pressure and weight of prisoners' experiences or the demands and constraints on prison staff' (Sloan & Drake, 2013, p. 25 [emphasis in original]). Yet, prison ethnographers do report their encounters with various vulnerabilities to harm: in emotional terms, 'the prison as a geographical and interpersonal space is emotionally demanding in and of itself (Drake & Harvey, 2014, p. 497); in existential terms, 'the prison is a "soul-sapping" institution, and no decent researcher can dodge the consequences' (Liebling, 2014, p. 482); and, in physiological terms, one researcher said to be 'ordinarily of good health', has experienced 'colds, flus, ulcers, and serious throat and bronchial infections, particularly when conducting research in high security prison settings' (Sloan & Drake, 2013, p. 35).

The point in all of this, again, is not to suggest that prisoners are not vulnerable human beings. Nor is it the case that researchers are ultimately 'as vulnerable as prisoners' in most meaningful senses of the word. Researchers are clearly privileged in relation to the imprisoned. The point is that vulnerability is not a straightforward 'thing' that is universally wielded by researchers over the imprisoned, at least not as the conventional narrative of research ethics abuses would have it. More importantly, I would argue, the non-dialectical, categorical approach to vulnerability currently enforced by IRBs has socio-spatially defined vulnerability in terms that do little to allay (or help mitigate) the inevitable anxieties associated with the field (cf. Jewkes, 2011). Instead, I am arguing, and I believe that contemporary prison ethnographers are implicitly arguing as well, that various forms of relational vulnerability characterize imprisonment research. As I now seek to illustrate, this is true within the prison, but also beyond.

Case study and discussion: finding and (re)placing vulnerability in practice

Methodological approach and the research design and review process

I worked with 48 former prisoners while conducting my dissertation research in Athens, Georgia, during the summers of 2008 and 2009 (Mitchelson, 2013). In the most general terms, my goal was to document and analyse the complicated spatial logics, practices, and outcomes that are associated with contemporary imprisonment. I was particularly interested in the relationship between imprisonment and the political representation(s) of prisoners in the decennial census (Lotke & Wagner, 2005; Tilove, 2002), and the urban population mobilities that imprisonment produces (Mitchelson, 2012). My formal interview protocol consisted of 26 questions: four functional questions about participants' journey through the criminal justice system (including jail(s) and prison(s)); eight about the census and counting; five about concepts of home; three about concepts of community; and six about concepts of race and ethnicity.

I also hoped to learn about the many things that I undoubtedly did not know about the geographies of imprisonment – things that I had yet to read, imagine, or experience myself. In turn, I utilized a semi-structured interview method that enabled me to engage with both my expectations and interests alongside emic themes as they emerged (Crang, 2002; Phillips & Earle, 2010). In total, I conducted 50 semi-structured interviews, which included follow-up interviews with 2 of the 48 participants. I used an unobtrusive digital voice recorder during the interviews in 49 of the interviews; in one instance, a participant requested that I not record the interview and I took notes by hand instead. Interviews were transcribed and triangulated with official records, as well as my personal field notes and research journal (Janesick, 1999, 2000).

I partnered with the Athens Justice Project (AJP), a local non-profit organization that engaged the legal, counselling and employment needs of indigent criminal offenders in the area. I initially approached the AJP's executive director for advice on how I might proceed in establishing relationships with local former prisoners, and a mutually beneficial relationship quickly developed. Staff members of the AJP included social workers, staff attorneys and interns; many of whom helped me work through early drafts of my interview protocol and recruitment materials in different ways. I eventually rented AJP office space, which was centrally located in the urban region, and, we believed, would establish an environment that was both welcoming and professional, and might help to mitigate some of the power dynamics inherent to the research process through its relative neutrality as a meeting place (Elwood & Martin, 2000). Roughly one third of the interviews I conducted were arranged in part by AJP employees; two thirds of the participants were snowballed from recruitment materials and informal word of mouth. I remunerated former prisoners for sharing their experiences – their expertise – with me, at an IRB-approved price-level ($50 per interview) established after consulting with area parole officers, AJP associates and potential participants.

In preparation for the field in which my participants and I would work, I had of course come across a large body of literature on vulnerability regarding the prison. And, in my case, I was told that I was working with people who my university formally and technically considered to be 'vulnerable participants'. I knew this in part because of item number nine on my university's Institutional Review Board (IRB) application, which was located between an explanation of how 'legally effective informed consent [would] be obtained from all research participants' and a binary check box (check Yes or No) for 'Illegal Activities'. Working through item number nine – unambiguously checked 'Yes' in my case under the heading 'VULNERABLE PARTICIPANTS' – with my advisory committee and the review board made my participants' vulnerability a point of special emphasis as my research design and protocol were vetted. It was probably 'the' point of emphasis.

The questions asked and addressed under the vulnerable participants heading in item number nine of my IRB application – about my participants' consent/assent to work with me, minimizing risks to them, and mitigating potential coercions of them during our research process, for example – were all crucial questions, no doubt. But, I would soon learn, they only indicated part of a bigger set of lessons regarding vulnerability and the research process. I turn now to the often unexpected ways that the research process unfolded. Although my experience by no means exhausts the lessons regarding vulnerability that will apply to a given research project, my hope is that there may be some suggestions or experiences that are of benefit to those considering or conducting research involving participants in a context of imprisonment.

Telling violence: 'I cut him just about all to pieces', George says

I was all cleared and ready to learn something, but what did I know at that point? Going in, I knew that a trip to prison can be a devastating fork in the road and that my participants were categorically disadvantaged by just about every conceivable social metric, from income to lifespan, simply because they had been to prison. I knew that there was a brutal legacy of ethical and callously violent physical abuses involving prisoner research (above). Emotionally, I remember feeling a sense of obligation to protect all of the vulnerable participants in/from my research process (Laliberte & Schurr, 2015). If I adopted the narrative of 'research subjects as victims and researchers as oppressors' then I could take comfort in casting myself as a benevolent protector of some sort, couldn't I? Sure, I wasn't wielding cholera or plutonium injections, but I was at least potentially dangerous. Wasn't that at least part of the take-home in reviewing the atrocities that prompted the Nuremberg Code or the Belmont Report?

There were clearly structurally and historically lop-sided power dynamics to contend with as my participants and I co-produced the research process. I was not in close contact with any past or present prisoners when I started fieldwork, and that fact alone suggests a great deal about the various privileges that I embody and experience on a given day. My participants could easily be misunderstood and, in turn, misrepresented. From my relatively privileged vantage point, I might not ever become aware of these distortions. There was going to be a lot of power on the table, so to speak. I had quite a lot of it. I couldn't undo the long legacies of structural and interpersonal violence that are part of our world; I could make things a bit worse. In particular, I was concerned that I might (unwittingly) impose epistemological acts of violence of my own, and simply extract convenient information from interviews, rendering the narratives as an academic commodity to the ends of personal and professional gain (Tuck & Yang, 2014).

I remember thinking: I have to protect my participants from this well-intentioned researcher and his research process, to which they are clearly vulnerable. And maybe that's all there is to it. All of the training and all of the signs seem to point to a particularly dangerous and immediate 'vulnerability' that is flowing in one, discrete direction: from me, onto my participants.

So, I am clearly carrying a lot of 'baggage' around with me in the field – what amounts to ethical and methodological insecurity, and practical inexperience – when I meet George.

George is the eldest of the folks with whom I worked – he is more than twice the average age of my participants. He speaks slowly and quietly. I sense a kindness in much of what he has to say, and a quaintness in the way that he describes the people and places upon which he reflects. In general terms, George's autobiographic narrative is a sad and interesting story of addiction, family, urban-to-rural migration in the early twentieth century south-eastern United States, the modern history of Atlanta, and crime and punishment in that city. When he speaks of his first incarceration – at age fourteen – the details are loose. 'I had a little trouble', he tells me, 'and I did a little time'. At other points in our interview, he talks about getting locked up for 'getting into little things, here and there'.

My protocol doesn't ask for, and my research questions don't require, any pre-incarceration specifics. I am committed to hearing what my participants want to tell me, but 'Tell me what happened' isn't really part of the script. For this work, I am happy enough to leave any details regarding what I am calling 'the crime part' alone; vague generalities are just fine by me. But George doesn't share all of his experiences with me in general terms.

Around the 10-min mark of our interview, George is telling me about bumping into a former acquaintance of his on the street. It strikes me as an everyday encounter. George is just on his way to the store. But things in the narrative take an abrupt turn here, although George's tone and intensity don't shift, and George says, 'This fella, he grabbed for my back pocket'.

'At that time, I carried a switchblade knife', George tells me. 'I took it out of my back pocket. I cut him here'. And George illustrates with his right hand. He moves it as if he is holding the knife and slashing my throat on both the left and right sides, quickly. He makes a soft whistling noise with both slashes. 'And here'. This time George raises his right arm and slashes down forcefully, in front of my chest, and then comes back across my stomach. He moves so quickly that I can't back away or react to his movements. I feel a push of wind as his hand swings past my face.

'I cut him just about all to pieces', George says.

George tells me that the man holds on for three months and then dies. George tells me, 'I got probation. I was so happy … I got off that time'.

A series of general descriptions of violent crime follow, the details of which are often hard to piece together. Several people are violently injured, and some of them die, in George's recollections. His next stories are about two women with whom he has physically violent experiences, after which he does time – at least two multi-year sentences in state penitentiaries. It is often unclear exactly how (or if) George fits into the violence he describes. For example, he tells me of a woman he once knew, 'I never will forget her. God bless her soul. She's dead and gone now'. He tells me about a happy window of time during which the two lived together. And then George tells me, 'There was evidence – evidence I had beat her around the face with my fists. So, they said, "George, we're gonna lock you up this time"'.

After describing his final incarceration, George tells stories of his recovery and his faith. These stories involve much more time and energy for George than had the descriptions of violence and incarceration in his past. And then, nearly 30 minutes after his unprompted recollections started, George claps his hands and asks, 'So, do you have any questions for me?' We hadn't even started with the protocol.

We work through my questions. We wrap the interview up, and have run longer than I'd anticipated. George and I shake hands, and he walks out of the office. Another participant is waiting for me, so I have to reset the interview process – paperwork and protocol – immediately. I hurriedly scribble down two words in my field notes before turning the page on my interview with George. Both of them are expletives.

Discovering, thinking and feeling through vulnerability

I have not seen or heard from George since that day, but I have reread the interview transcript and listened to the recording many times. When I play it back, I don't get from George that he is consciously trying to intimidate me when he imitates his lethal movements or describes the homicide. I don't get from George that he is a man prone to – perhaps even capable of – committing such acts of violence at this stage of his life. Yet, I had to at least consider the possibility that I had just spent time in close quarters with a habitually violent human being who posed a legitimate threat to my well-being, and the well-being of others. It had to be on the table as a possibility, however remote. And, in hindsight, I think that intellectually accepting the possibility of my own bodily vulnerability to violence actually helped me to be a more intellectually and emotionally 'present' participant.

I am not George's judge now, and I don't hear signs that I took that role on during our time together – however visceral my internal reactions may have been (and they were), I honoured my commitment to let the story unfold without inserting my assumptions or judgements. When I listen back, I sound like a young researcher who is truly open to hearing what his participant has to say, and I am very proud of that. At the same time, I had no idea – absolutely no idea – how much energy that kind of openness and attention could consume.

Imprisonment doesn't stop or end at the prison's walls. Imprisonment gets built into the experiences, environment and choices that an individual prisoner takes 'in' with them. Imprisonment gets borne back 'out' in many painful ways. My work with former prisoners taught me a lot about resilience and triumph, but nobody had a happy prison story. Listening to human beings describe many of the very worst experiences of their lives often *hurts*. A lot. And while that is completely unsurprising from a common sense or experiential perspective, it was completely unmentioned in the hours of Collaborative Institutional Training

Initiative (CITI) training or IRB review and revision that were required of me. Importantly, that pain would have likely been very similar, if not identical, had the interview taken place within a prison. That pain was not location-specific, it was relationally produced.

The pain that I experienced was something that I was fortunate to work through and process with great friends and colleagues (Jewkes, 2011; Kohl & McCutcheon, 2014). That pain, I would argue, suggests that I was risking vulnerability. Which means, I think, that I was doing at least some things effectively as a qualitative researcher and participant. I still haven't fully processed the way that interview feels (Laliberte & Schurr, 2015). But, I am aware that it decisively ended the possibility of thinking about the vulnerability of 'my participants' and replaced that notion with the vulnerabilities produced through 'our participation' in engaged research.

Placing other relational vulnerabilities

In recognizing myself as a vulnerable participant in the research process, I joined a group of people who had alerted me to their own sense(s) of vulnerability. Those closest to me at the time – family, friends and colleagues – also expressed concerns about their (and my) proximity to a group of participants about whom they had multiple, complex, and sometimes contradictory notions. More than one person expressed concerns about feeling vulnerable. So, establishing healthy boundaries as I would hope to in any relationship (whether premised on research or not) became an important factor of consideration in mitigating perceived vulnerabilities in all of this in at least two ways. The concerns raised by those closest to me primarily centred on money and accessibility.

The first person who raised concerns about their vulnerability to me was a member of the AJP. Remuneration was paid in cash following each interview, after participants completed a standard institutional payment form. As was the case when I interviewed George, I conducted many interviews on a tight timeline, and I often conducted several interviews on the same day. Functionally, that meant that I often walked into the AJP office with a lot of cash – sometimes several hundred dollars. By all accounts and appearances, the AJP employees and clients had trusting, healthy relationships. I observed a lot of mutual respect in their office, and felt comfortable with my belongings and the cash. But, I was reminded that while my research may have started with their clientele, it rather quickly snowballed out of that group, and attracted other people. Some were strangers, and some were people with whom the AJP had chosen not to partner for various reasons, or had severed relations. This sent me back to the drawing board – no one at the IRB level of review had considered the ramifications of hauling this cash around. Neither had I.

I believed that I could internalize this perceived vulnerability, and made efforts to that end. I started making explicit statements about the cash – that the cash came in with me, and it left with me. I also explicitly clarified that I was not an AJP employee, and that there was no other cash kept on premises. I always kept all of the cash in the same envelope, in the same blue binder, which I clearly showed to each of the participants as I explained the cash procedures and we filled out the attendant paperwork.

Before cash entered the research process – conversationally, or, in these deliberate acts of performative 'security' – I worked to establish a mutually respectful exchange. Gratitude was a legitimate, crucial part of our interactions. I made sure that my participants knew that I was learning from them – things that I could not know without them – and that I hugely valued, and prized, their contribution to my work. I thanked them at the beginning, throughout, and after each interview, for sharing many of the worst experiences of their lives with me. This earnest gratitude seemed to engender a spirit of kindness and reciprocity. For example, I was hugged often; not always, but often, at the conclusion of an interview.

I communicated these changes and commitments as assurances to the AJP employee who had approached me, and it seemed to quell the suspicion/perceived risk. We proceeded apace without slowing or pausing the research process. Ultimately, no one ever called the cash into question again. I was never approached by anyone for cash before or after my time in the office.

My family members, in particular, expressed concerns regarding accessibility. In working through my responses to their concerns, I used a disposable cell phone (a 'burner') to work with scheduling participants. The phone number was initially published and distributed amongst AJP clients and I know at least anecdotally that it circulated well beyond that group. This phone created a few layers of distance and security that were important in my case.

First, folks calling in were assured autonomy and anonymity. They could call from anywhere. They could feel me out before scheduling an interview, weigh the information presented to them, and make a choice about whether or not they wanted to schedule an interview. They could schedule an interview and simply not show if they reconsidered and an interview ultimately failed to meet their needs. The phone also came in handy on a couple of occasions, as participants were able to use my phone to schedule rides and make other arrangements, as well as (on two separate occasions) call to make preliminary connections with additional participants.

The phone was also of clear practical benefit to me. For example, there was a voice mail function, which effectively enabled potential participants to contact me at any time. The phone also allowed me the mobility that all cell phones afford their owners, but, unlike many service plans, the ownership and activation of this phone permitted me to put a finite beginning and ending on that phase of research. In terms of my own autonomy and relative anonymity, the phone afforded me some level of privacy. It was not directly linked to me and/or my home address. Yet, the phone gave me a means by which I could stay connected *from* home without participants connecting *to* my home, which was expressed as a need of my family's.

The phone signals the crucial role of distance in imprisonment research. It afforded me and my participants a level of mutual accessibility that did not appear to be constrained by income or social status, time or location. It put pragmatic boundaries around that phase of the research. People who wanted to stay in touch with me following the study shared contact information; people who didn't did not.

Persistent vulnerability

I have reunited with several participants in social settings after the formal research process had concluded. And I have been able to hear of triumphant healing and recovery processes unfolding for several of the folks with whom I worked, many of whom were up against brutal odds when we first met. Those are wonderful experiences. But they aren't exhaustive of the residual vulnerabilities – positive and negative – that a research process can generate. My experience with a participant named KP is suggestive.

It always seemed important to KP that I understand clearly that nobody had seen his incarceration coming, and it was a part of his life that he'd left behind. We laughed several times during the interview, and seem to have established a friendly rapport when I review the transcript and tape.

At the interview's close, KP tells me,

> This is probably the most I've talked about my prison time in years. I'm not ashamed of anything I've done. I'm not proud of it, but I'm not ashamed of it. If I had it to do over, I would probably do things differently. But there's no sense living in shame because I went to prison.

That is a point on which we readily agree, and the final seconds of the interview are mutually amicable.

I bumped into KPon several occasions after our interview. The first time was only a day or two after the interview. We passed each other on the street, and I remembered him well. He nodded when I greeted him and moved on. We went our separate ways, and I didn't think much of it.

I was at a departmental happy hour, several months later, the next time I saw KP. It was a great night, characterized by a lot of enthusiastic and supportive camaraderie; someone had successfully defended their dissertation proposal, if memory serves. I remember being in a great mood when KP walked in. We recognized but didn't acknowledge one another beyond a nod. KP was with friends, and as best I could tell they seemed to all be in a great mood, too. It wasn't a big place, and I didn't want to intrude on their time. I went about my business but looked back over my shoulder just in time to see KP regrouping his friends. They left.

It was still a great night, but I remember pondering KP's seemingly abrupt departure for several days. Some of the possible explanations for his departure merit close consideration, several of which suggest that relational vulnerability is circuital. First, there is no getting around the fact that KP had disclosed things to me in a particular, ostensibly secure, environment. In my view, the best interview would be defined by mutual vulnerability, because the researcher would be radically open to the listening process and the interviewed participant would be freely (re)opening their experiential knowledge and insights. Maybe 'going back' had hurt KP.

Or, maybe I had unknowingly brought judgemental thinking or criticism to the interview as a listener. Maybe I had done something unintentionally offensive, or reminded him of someone with whom he'd had negative experiences. Did I owe KP an apology for an unintentional offense? I went back to the tape the next morning, listening for possible offenses on my part, and didn't catch anything (although that doesn't necessarily mean that it wasn't there).

Maybe there was a more innocuous explanation. If our mutual friends started socializing that evening, how would either of us respond to the question, 'How do you guys know each other?' Maybe KP just didn't want to deal with that.

Or, it might have all been a coincidence – maybe they just had to go.

It was my job to interview KP and learn what he could teach me. It was not my job to be his friend any more or less than the rest of my acquaintances. So, although I did make every effort to be friendly, I decided to leave it all be. KP had my contact information if he needed to inform me of an offense I'd committed.

I saw him last month, and he nodded. We passed our ways and moved on. My working hypothesis is that this distance is likely for the best. We remain vulnerable to one another, because we co-produced a meaningful and informative interview process together.

Location is neither irrelevant to, nor determinant of, vulnerability in any of this. The explicit premise for our engagement with one another was KP's incarceration – the time during his life when he was located within a prison's walls. His vulnerability to 'his past' and his emotional relationship to it, through his conversation with me (if I am rightly onto discomfort on KP's part in this anecdote), are not necessarily determined by his imprisonment, however. KP could be equally vulnerable to 'his past' while still confined, had the interview taken place in a prison and had we seen one another after his release. More interestingly, perhaps, KP might have responded in exactly the same way if our interview had been premised on any difficult time during his life in any setting. Because location's role in the vulnerabilities at work in this research is so ambiguous, categorically treating research participants as vulnerable solely on the basis of their location seems highly insufficient to the task of conducting research in mutually safe, healthy and meaningful ways.

Summary and concluding thoughts

In this article, I have worked to closely examine the ethical, methodological and practical implications of vulnerability in my work with former prisoners. Although much of the discussion documented here takes an anecdotal form, it is my hope that some of what I have reported may be of use to others in shaping their own research process. Or, that others may insightfully propose alternative strategies and analysis for working through vulnerability.

I have sought to prove two points in support of my summative argument: that the politically reactive legal designation of 'vulnerable populations' may miss its mark in identifying actually existing vulnerabilities, and, ostensibly, protecting participants from undue harm during the research process. My first point is that vulnerability is an inherently relational component of any human-geographic research project. Rather than a 'thing' that might be wielded, I have worked to document vulnerability as an unfolding process that is shared by researchers and other, often distant or indirectly associated, participants. George, KP, and I risked vulnerability by engaging one another in earnest. And while I would never argue that we engaged one another on an equal footing (I clearly brought a litany of relative privileges to the research process), the co-production of our interviews' situated knowledge was shared nonetheless. My second point is that vulnerability – due to its relational, processual nature – is poorly defined in exclusively locational terms. Prisons are violent and coercive settings; those within them are vulnerable. However, imprisonment is a process that includes people and places well beyond the prisons walls, and so too should any attendant definition of vulnerability that is premised on the prison.

In closing, I want to signal three avenues of future inquiry that can contribute meaningfully to our understanding of vulnerability, particularly in the case of imprisonment. The first is a more in-depth historical accounting of 'vulnerability' as a legal category. The fact that the Nuremberg Code resulted from a US prosecution, after which seven prisoners (i.e. Nazi doctors) were put to death, and after which prison research in the United States then actually flourished instead of abating, merits close analysis. The work of Harkness(1996), which examines the political process through which this apparent contradiction unfolded, is particularly intriguing. Understanding the work that 'vulnerability' was designed to do may prove instructive for recreating a more ethically, methodologically and practically meaningful set of protections against legitimate harms associated with research.

The second avenue for further inquiry regards vulnerability in the context of the prison, specifically. The relative inaccessibility of prisons as research sites has been well documented, and a practical concern is that bureaucratic and security concerns will result in the addition of months – if not years – to a given research process if it takes place (in whole, or in part) within a prison (Cislo & Trestman, 2013; Wacquant, 2002). Yet, this inaccessibility comes at a crucial point in the prison's history. There were fewer than 200,000 people in state and federal prisons when the Nuremberg Code was issued. That number has grown by more than 800% in the intervening years. Further, several legal and medical scholars warn that the ostensibly well-intended measures enforced in the name of protecting 'vulnerable populations' may be doing more harm than good at this time (Hoffman, 2000; Levine et al., 2004; McDermott, 2013). So, while research on imprisonment presents a host of ethical, methodological and practical challenges to those who are willing to risk relational vulnerability, it presents a host of empirical and philosophical questions that have yet to be asked.

The third avenue would connect this prison-themed work on vulnerability with broader debates about vulnerability and research ethics, and IRB oversight, in particular. Geographers have already contributed substantive critiques of this approach to the regulation of research ethics (e.g. Bradley, 2007; Martin & Inwood, 2012). Schneider's recent monograph (2015) – entitled *The Censor's Hand: The Misregulation of Human-Subjects Research* – is scathing and extensive in its criticism of the current framework. It is beyond the scope of this paper address the question of IRB oversight in general. However, in my view, the case of its imprisonment research ethics oversight is problematic for at least two reasons, both of which have been touched on above. First, the historical origins of an IRB's working concepts (e.g. 'vulnerable populations') and current federal regulations are highly problematic. The Belmont Report's treatment of its 'three fundamental ethical principles' – Respect for Persons, Beneficence, and Justice – is not 'a diagnostic tool' (Cislo & Trestman, 2013, p. 306). It is not designed to empirically measure or assess the ethical status of a given research initiative. In this sense, the report perhaps 'encourages more questions than it seeks to answer' (Kalmbach & Lyons, 2003, p. 677). An ambiguously worded political reaction to research scandals that are represented by geographic imaginaries is likely always going to miss its mark in affording meaningful protections to specific people involved with specific research projects.

When it is produced and overseen as a categorical construct, the federal government and IRB's framework for research ethics fixes vulnerability in space and time. When it is 'fixed' in this way, difference and agency are assumed away and prisoners become universally 'vulnerable' as a population. Whether vulnerability must be regulated in research or not is a larger ethical question with sweeping ramifications. However, if the current framework is to undergo reform, perhaps vulnerability can be more usefully and meaningfully considered at the level of research process, as opposed to location, in the context of imprisonment and other research efforts in the humanities and social sciences. This would have at least two benefits. First, it would result in IRBs considering humanities and social science research on their own terms, as opposed to adopting a set of biomedically-rooted categories and criteria (Burr & Reynolds, 2010). Second, and far more pressingly, this would enable a process-based (i.e. as opposed to locationally-defined population-based) evaluation of vulnerability in relational terms. I am hopeful that such a change might indeed help to generate 'more interesting and honest knowledge' (Jewkes, 2011, p. 72) of imprisonment and those impacted by it.

Disclosure statement

No potential conflict of interest was reported by the author.

Funding

This work was supported by the National Science Foundation [grant number 0727443].

References

Baer, L. D. & Ravneberg, B. (2008). The outside and inside in Norwegian and English prisons. *Geografiska Annaler, Series B: Human Geography, 90*, 205–216.

Bosworth, M., Hoyle, C., & Dempsey, M. M. (2011). Researching trafficked women: On institutional resistance and the limits to feminist reflexivity. *Qualitative Inquiry, 17*, 769–779.

Bradley, M. (2007). Silenced for their own protection: How the IRB marginalizes those it feigns to protect. *ACME: An International E-Journal for Critical Geographies, 6*, 339–349.

Brenner, N., & Theodore, N. (2002). Cities and the geographies of "actually existing" neoliberalism. *Antipode, 34*, 349–379.

Burr, J., & Reynolds, P. (2010). The wrong paradigm? Social research and the predicates of ethical scrutiny. *Research Ethics Review, 6*, 128–133.

Butler, J. (2004). *Precarious life*. New York, NY: Verso.

Cislo, A. M., & Trestman, R. (2013). Challenges and solutions for conducting research in correctional settings: The U.S. experience. *International Journal of Law and Psychiatry, 36*, 304–310.

Comfort, N. (2009). The prisoner as model organism: Malaria research at Statesville Penitentiary. *Studies in History and Philosophy of Biological and Biomedical Sciences, 40*, 190–203.

Committee Appointed by Governer Dwight H. Green of Illinois. (1948). Ethics governing the service of prisoners as subjects in medical experiments. *Journal of the American Medical Association (JAMA), 136*, 457–458.

Crang, M. (2002). Qualitative methods: The new orthodoxy? *Progress in Human Geography, 26*, 647–655.

Crewe, B. (2014). Not looking hard enough: Masculinity, Emotion, and Prison Research. *Qualitative Inquiry, 20*, 392–403.

Crewe, B., Warr, J., Bennett, P., & Smith, A. (2014). The emotional geography of prison life. *Theoretical Criminology, 18*, 56–74.

Drake, D. H., & Harvey, J. (2014). Performing the role of ethnographer: Processing and managing the emotional dimensions of prison research. *International Journal of Social Research Methodology, 17*, 489–501.

Elwood, S. A., & Martin, D. G. (2000). "Placing" interviews: Location and scales of power in qualitative research. *Professional Geographer, 52*, 649–657.

Few-Demo, A. L., & Arditti, J. A. (2014). Relational vulnerabilities of incarcerated and reentry mothers: Therapeutic implications. *International Journal of Offender Therapy and Comparative Criminology, 58*, 1297–1320.

Few, A. L., & Rosen, K. H. (2005). Victims of chronic dating violence: How women's vulnerabilities link to their decisions to stay. *Family Relations, 54*, 265–279.

Gilmore, R. W. (2007). *Golden gulag: Prisons, surplus, crisis, and opposition in globalizing California*. Berkeley: University of California Press.

Gilmore, R. W. (2008). Forgotten places and the seeds of grassroots planning. In C. R. Hale (Ed.), *Engaging contradictions: Theory, politics, and methods of activist scholarship* (pp. 141–162). Berkeley: University of California Press.

Harkness, J. M. (1996). Nuremberg and the issue of wartime experiments on US prisoners. *JAMA, 276*, 1672–1675.

Heller, J. (1972, July 26). Syphilis victims in U.S. study went untreated for 40 years. *New York Times*.

Hoffman, S. (2000). Beneficial and unusual punishment: An argument in support of prisoner participation in clinical trials. *Indiana Law Review, 33*, 475–515.

Hornblum, A. M. (1997). They were cheap and available: Prisoners as research subjects in twentieth century America. *BMJ, 315*, 1437–1441.
Huijsmans, R. (2012). Beyond compartmentalization: A relational approach towards agency and vulnerability of young migrants. In A. Orgocka & C. Clark-Kazak (Eds.), *Independent child migration-insights into agency, vulnerability and structure: New directions for child and adolescent development, number 136* (pp. 29–45). Hoboken, NJ: Wiley Periodicals, Inc.
Institute of Medicine of the National Academies. (2006). *Ethical considerations for research involving prisoners*. Washington, DC: The National Academies Press.
Janesick, V. J. (1999). A journal about journal writing as a qualitative research technique: History, issues, and reflections. *Qualitative Inquiry, 5*, 505–524.
Janesick, V. J. (2000). The choreography of qualitative research design: Minuets, improvisations, and crystallization. In N. K. Denzin & Y. S. Lincoln (Eds.), *Handbook of qualitative research* (pp. 379–400). Thousand Oaks, CA: Sage.
Jewkes, Y. (2011). Autoethnography and emotion as intellectual resources: Doing prison research differently. *Qualitative Inquiry, 18*, 63–75.
Jewkes, Y. (2014). Special section editorial: An introduction to "doing prison research differently". *Qualitative Inquiry, 20*, 387–391.
Jones, K. T., & Popke, E. J. (2010). Re-envisioning the city: Lefebvre, Hope VI, and the neoliberalization of urban space. *Urban Geography, 31*, 114–133.
Kabeer, N. (2014). *Violence against women as 'relational' vulnerability: Engendering the sustainable human development agenda*. New York, NY: UNDP Human Development Report Office.
Kabeer, N., Khawar, M., & Sayeed, A. (2010). Beyond risk management: Vulnerability, social protection and citizenship in Pakistan. *Journal of International Development, 22*, 1–19.
Kalmbach, K. C., & Lyons, P. M. (2003). Ethical and legal standards for research in prisons. *Behavioral Sciences and the Law, 21*, 671–686.
Katz, C. (1994). Playing the field: Questions of fieldwork in geography. *Professional Geographer, 46*, 67–72.
Kohl, E., & McCutcheon, P. (2014). Kitchen table reflexivity: Negotiating positionality through everyday talk. *Gender, Place and Culture: A Journal of Feminist Geography, 22*, 747–763.
Laliberte, N., & Schurr, C. (2015). The stickiness of emotions in the field: Complicating feminist methodologies. *Gender, Place and Culture: A Journal of Feminist Geography, 23*, 72–78.
Lerner, B. H. (2007). Subjects or Objects? Prisoners and Human Experimentation. *The New England Journal of Medicine, 356*, 1806–1807.
Levine, C., Faden, R., Grady, C., Hammerschmidt, D., Eckenwiler, L., & Sugarman, J. (2004). The limitations of "vulnerability" as a protection for human research participants. *The American Journal of Bioethics, 4*, 44–49.
Liebling, A. (2014). Postscript: Integrity and emotion in prisons research. *Qualitative Inquiry, 20*, 481–486.
Lotke, E., & Wagner, P. (2005). Prisoners of the census: Electoral and financial consequences of counting prisoners where they go, not where they come from. *Pace Law Review, 24*, 587–607.
Martin, D. G., & Inwood, J. (2012). Subjectivity, power, and the IRB. *The Professional Geographer, 64*, 7–15.
Martin, L., & Mitchelson, M. L. (2009). Geographies of detention and imprisonment: Interrogating spatial practices of confinement, discipline, law, and state power. *Geography Compass, 3*, 459–477.
McCarthy, C. (1989). Experimentation on prisoners: The inadequacy of voluntary consent. *New England Journal on Criminal and Civil Confinement, 15*, 55–80.
McDermott, B. E. (2013). Coercion in research: Are prisoners the only vulnerable population? *The Journal of American Academy of Psychiatry and the Law, 41*, 8–13.
McKittrick, K. (2012). On plantations, prisons, and a black sense of place. *Social & Cultural Geography, 12*, 947–963.
Merrifield, A. (1993). Place and space: A Lefebvrian reconciliation. *Transactions of the Institute of British Geographers, 18*, 516–531.
Mitchelson, M. L. (2012). The urban geography of prisons: Mapping the city's 'other' gated community. *Urban Geography, 33*, 147–157.
Mitchelson, M. L. (2013). Up the river (from home): Where does the prisoner 'count' on census day? In D. Moran, N. Gill, & D. Conlon (Eds.), *Carceral spaces* (pp. 77–92). Surrey: Ashgate.

Mitchelson, M. L. (2014). The production of bedspace: Prison privatization and abstract space. *Geographica Helvetica, 69*, 325–333.

Moran, D. (2013). Leaving behind the 'total institution'? Teeth, transcarceral spaces and (re)inscription of the formerly incarcerated body. *Gender, Place and Culture: A Journal of Feminist Geography, 21*, 35–51.

Petersilia, J. (2001). When prisoners return to the community: Political, economic, and social consequences. *Corrections Management Quarterly, 5*, 1–10.

Phillips, C., & Earle, R. (2010). Reading Difference Differently? *British Journal of Criminology, 50*, 360–378.

Reiter, K. (2014). Making windows in walls: Strategies for prison research. *Qualitative Inquiry, 20*, 417–428.

Reverby, S. M. (2013). *Examining Tuskegee: The infamous syphilis study and its legacy*. Chapel Hill: The University of North Carolina Press.

Rowe, A. (2014). Situating the self in prison research: Power, identity, and epistemology. *Qualitative Inquiry, 20*, 404–416.

Schneider, C. E. (2015). *The censor's hand: The misregulation of human-subjects research*. Cambridge, MA: The MIT Press.

Sloan, J., & Drake, D. H. (2013). Emotional engagements: On sinking and swimming in prison research and ethnography. *Criminal Justice Matters, 91*, 24–25.

Soja, E. W. (1980). The socio-spatial dialectic. *Annals of the Association of American Geographers, 70*, 207–225.

Taylor, M. (2013). Climate change, relational vulnerability and human security: Rethinking sustainable adaptation in agrarian environments. *Climate and Development, 5*, 318–327.

Tilove, J. (2002). Minority prison inmates skew local populations as states redistrict. *Newhouse News Service*. Retrieved December 20, 2005, from http://www.newhousenews.com/archive/story1a031202.html

Travis, J. (2001). But they all come back: Rethinking prisoner reentry. *Corrections Management Quarterly, 5*, 23–33.

Tuck, E., & Yang, K. W. (2014). R-words: Refusing research. In D. Paris & M. T. Winn (Eds.), *Humanizing research: Decolonizing qualitative inquiry with youth and communities* (pp. 223–248). Los Angeles, CA: Sage.

Ugelvik, T. (2014a). *Power and resistance in prison: Doing time, doing freedom*. Basingstoke: Palgrave Macmillan.

Ugelvik, T. (2014b). Prison ethnography as lived experience: Notes from the diaries of a beginner let loose in Oslo prison. *Qualitative Inquiry, 20*, 471–480.

Wacquant, L. (2002). The curious eclipse of prison ethnography in the age of mass incarceration. *Ethnography, 3*, 371–397.

Watts, M. J., & Bohle, H. G. (1993). The space of vulnerability: The structure of hunger and famine. *Progress in Human Geography, 17*, 43–67.

Researching migration and enforcement in obscured places: practical, ethical and methodological challenges to fieldwork

Pauline Maillet, Alison Mountz and Keegan Williams

ABSTRACT

This paper explores practical, methodological and ethical concerns of research on and in sites where human migration is regulated. There, enforcement practices such as policing, interception and detention are transpiring, but often hidden from view. Access to data about these sites is highly restricted and partial. We draw on recent research about migration and enforcement at sea, in airport waiting zones and on islands where people are detained and have their mobility restricted. We discuss empirical issues to show the complexity of methodological decision-making amid asymmetrical power relations in distinct, yet related settings where human mobility is regulated and borders enforced. One of our aims is to question and complicate notions of vulnerability in places where people are categorized as 'vulnerable'. We argue that, despite the challenges and complexities detailed in this article, it is extremely important to do this kind of research. Conversations about this work, though difficult, are worth having and recording, so that they continue beyond these pages and researchers do not feel alone in struggling with these issues.

Recherche en migration et application des lois dans les lieux occultés: défis pratiques, éthiques et méthodologiques du travail sur le terrain

RÉSUMÉ

Cet article explore les inquiétudes pratiques, méthodologiques et éthiques des chercheurs sur et dans les sites où est règlementée la migration humaine. Là, les pratiques d'application des lois telles que le contrôle, l'interception et la détention se font jour mais sont souvent occultées. L'accès aux données au sujet de ces sites est extrêmement restreint et partial. Nous nous appuyons sur les recherches récentes au sujet de la migration et de l'application des lois en mer, dans les zones d'attente des aéroports et sur les îles où les gens sont détenus et où leur mobilité est restreinte. Nous abordons des questions empiriques pour montrer la complexité de prise de décision méthodologique au sein de relations de pouvoir asymétriques dans des cadres distincts et pourtant liés où la mobilité humaine est contrôlée et les frontières imposées. Un de nos buts est de remettre en question et de compliquer les notions de vulnérabilité dans des lieux où les gens sont catégorisés comme « vulnérables ». Nous postulons que, malgré les défis et les complexités exposés dans cet article, il est extrêmement important de faire ce genre de recherche. Cela vaut la peine, bien que ce soit difficile, d'avoir des conversations au sujet de ce travail et de les enregistrer, afin qu'elles puissent continuer au-delà de ces pages et que les chercheurs ne se sentent pas seuls à débattre ces questions.

Analizando la migración y la ejecución en lugares ocultos: desafíos prácticos, éticos y metodológicos para el trabajo de campo

RESUMEN
Este artículo explora intereses prácticos, metodológicos, y éticos de la investigación sobre y en sitios donde se regula la migración humana. Allí, tienen lugar prácticas de aplicación tales como la vigilancia, la intercepción y la detención, pero a menudo no están a la vista. El acceso a los datos acerca de estos sitios es altamente restringido y parcial. Este trabajo se basa en investigaciones recientes sobre migración y ejecución en el mar, en zonas de espera del aeropuerto, y en islas cuando las personas son detenidas y su movilidad es restringida. Se discuten problemas empíricos para mostrar la complejidad de la toma de decisiones metodológicas en medio de relaciones de poder asimétricas en los entornos distintos, pero relacionados, donde se regula la movilidad humana y se imponen límites. Uno de los objetivos es cuestionar y complicar las nociones de vulnerabilidad en lugares donde la gente es categorizada como 'vulnerable'. Se argumenta que, a pesar de los desafíos y complejidades detallados en este artículo, es muy importante hacer este tipo de investigación. Aunque es difícil, vale la pena tener y grabar conversaciones sobre este trabajo, para que continúen más allá de estas páginas y los investigadores no se sientan solos en la lucha en relación a estos temas.

1. Introduction

This paper explores practical, methodological and ethical concerns of research on and in sites of human migration where enforcement practices such as policing, interception and detention are transpiring, but often hidden from view. In these sites, authorities restrict human mobility and render partial most data available to the public. While it is easy for researchers to become frustrated or even paralyzed when confronting the challenges to accessing this kind of information, we suggest that it is important to press on in spite of and, indeed, because of these challenges. But key to careful and successful research in these politicized locations is continuous conversation about challenges and possible ways forward. In order to promote and provoke such conversations, we discuss empirical issues to show the complexity of methodological decision-making amid asymmetrical power relations in distinct settings where human mobility is regulated and borders enforced. We discuss, for example, participant-observation in secure facilities under surveillance; projects *about* detention, migration and asylum not premised solely on *entering* facilities; whether or not to share potentially damaging revelations uncovered during fieldwork; and whether to conduct interviews inside facilities. As we show in the case of 'missing data', confined spaces at the border can be electronic as well as physical or legal. By discussing decisions about when, where, how and with whom to conduct research on migration and detention in highly politicized circumstances, the article aims to prompt rather than neglect conversations about research in politically charged arenas.

We argue that, despite its difficulties and complexities, it is extremely important to do this kind of research. As Hyndman (2001) notes, 'there is value in working through the messiness, engaging in fieldwork in a careful manner, rather than writing it off as too fraught with difficulties and dangers. Imperfect engagement is better than no engagement, or a paralyzing angst' (p. 265). We agree and argue that these conversations, though difficult, are worth having and sharing, so that they continue beyond these pages and researchers do not feel alone in their struggles. Isolation, in fact, motivated us to co-author this article, and we suggest that it links researchers to the 'vulnerable' whom they contact in their research.

In order to develop our argument, we draw on recent research about migration and enforcement at sea, in airport waiting zones and on islands where people are detained and have their mobility restricted. This work involved accessing, documenting and disseminating information about migration, asylum and detention in remote locations intentionally hidden from the view of publics, journalists, human rights monitors and researchers. In other words, this is information that state authorities often suppress.

One of our aims is to question and complicate notions of vulnerability. Many institutional bodies – such as university research ethics boards designed to protect human subjects – construct migrants, asylum seekers and detainees in these places as 'vulnerable'. At the same time, the same people are treated as security threats by authorities who, ironically, construct citizens as vulnerable. These shifting constructions of vulnerability intersect and collide with complex debates surrounding representation and voice. Whose voices are heard in research? Who has the power to represent? How is representation challenged or made impossible by the power relations and locations where research processes happen?

In this article, we do not necessarily contest migrants as vulnerable or threatening. Like everyone, they are a heterogeneous grouping, with individuals taking on each of these characteristics. Furthermore, people enter and exit varying degrees of vulnerability and precarity at different times and in different places. We problematize vulnerability by exploring different ways of conceptualizing migration, detention and those involved in a variety of capacities. Examples throughout the article unpack vulnerability. As we will demonstrate, migrants are not the only participants who may be rendered vulnerable by research (Behar, 1996; Bondi, 2005).

Language is also political and can contribute to vulnerability. As Bauder (2013) contends, 'language matters in public discourse and everyday exchange: terminology can imply causality, generate emotional responses and transmit symbolic meanings' (p. 2). He concurs with Barnett and Duvall (2005) who, in their analysis of productive power, note that discourse produces subjects with associated rights and responsibilities. Such classification has real consequences for people on the ground. For example, assignment to the categories of 'civilian' or 'combatant' translates into differential protection under international humanitarian law (Barnett & Duvall, 2005). Similarly, international law has established a specific legal regime of protection for asylum seekers and refugees.

As researchers, we grapple with terminology as we write: How should we refer to groups we study? Various institutions label them migrants, aliens, undocumented, unauthorized, non-admitted foreigners, detainees, third country nationals, asylum seekers, refugees and passengers. National and international law prescribe varying rights to these categories of people; public discourse similarly assigns value and devalues. Much of the enforcement we study works by controlling access to these rights, such as the right to seek asylum. Although people often do not identify with these categories, they may try to or become assigned as subjects as a consequence of state exclusion (Massey & Capoferro, 2008, pp. 259–265). Moreover, individuals change status at different stages of their lives or journeys, including periods of confinement. At sea, in airport waiting zones and in detention facilities, for example, people move frequently between 'non-admitted alien' and 'asylum-seeker'.

Activists and scholars resist use of the term 'illegal' as it implies either that a person has committed a crime or that personhood itself is somehow illegal (consider the slogan: 'No one is illegal'). Either way, the term triggers a negative emotional response. Bauder (2013) suggests, instead, the term 'illegalized immigrant' to convey the idea that people are *rendered* illegal. In this article, we recognize that language and categories directly affect the lives of people whom we study and are wary of the language of the state. Confronted by a dependency on documents produced by enforcement authorities, we are forced to critically analyse how they construct and deploy migrant categories. We simultaneously read, cite and produce migration research, which, like so much social science, tends to homogenize individual stories through categorization, sorting and counting. We do not claim to have solved these issues associated with terminology, but, instead, question and mindfully engage the proliferating panoply of migrant categories used in different contexts throughout our projects.

In the spirit of conversation, we proceed with dialogue on our collective reflections about these issues in three different arenas: island detention facilities, waiting zones at international airports and interceptions at sea. We explore and share ethical challenges in our own research projects with the hope that this will extend conversations among others facing similar issues. We began this dialogue with the following questions as starting points: What methods are appropriate when states hide information? What ethical issues arise throughout the research process, from data collection to dissemination of findings? How have we handled these issues, and why did we press on? Having found that existing literature inadequately addresses these questions, we continue by staging each section of our paper as an answer to the questions that *we* have about this work in which we have been engaged. Our hope is that others embarking on projects will find conversation about ethical, practical and methodological dimensions of research on migration and exclusion helpful.

2. Why are we having this conversation? An introduction to three projects

In this section, we introduce three research projects that led to this discussion, and briefly explore positionality therein. This enables us to write, where possible, in a collective voice while contextualizing examples used throughout the article. As researchers, we hold differentiated access to spaces and resources compared to migrants, NGOs or state actors. At the same time, we have experiences as migrants, activists and NGO workers, which inform our understanding of each project. Our work, therefore, reflects the process of negotiating our positions and ethical commitments in the face of complex issues. While being a researcher provides some institutional protections and resources, we will show that it also restricts access to information when studying confined spaces and wide-open places, presenting ethical and methodological challenges.

We describe the projects in chronological order. The first is the Island Detention Project on which Author 2 served as Principal Investigator, with qualitative field research conducted by a team of researchers working in three regions between 2010 and 2013. This project emerged from years of research on unauthorized, undocumented migration and asylum seeking, where Author 2 found that nation states were becoming more aggressive in pushing border enforcement offshore. Increasingly, those affected by this offshoring were not recorded in national statistics; instead, they were excluded offshore, prevented from entering, or included by their very exclusion. The project emerged from a desire to include the experiences and voices of asylum seekers and migrants marginalized through interception and detention in remote locations. There was a commitment to do this responsibly in ways that neither co-opted their stories nor made them more vulnerable, given their precarious legal situations. This often proved challenging.

Ethical limitations here related to people's vulnerability as asylum seekers and migrants confined in detention. Not only was their mobility restricted, but they were under surveillance and information gathered in these settings could not necessarily be protected or anonymized. As a result, researchers did not conduct formal interviews inside of facilities, but, rather, outside of them. Researchers also conducted participant-observation on the inside. They accomplished this by working, wherever possible, with established local visitors' groups, whether NGOs, legal representation for those on the inside or more loosely organized visitors working as volunteers. By working with people who had established trust and worked respectfully over time with people who were detained, they hoped to contribute to their cause and advocacy, rather than enter in potentially damaging ways into facilities. Through this participant-observation, facilities were conceptualized as networks, concentric zones, embedded simultaneously in local contexts, relationships, economies and global circuits of transnational social relations. Researchers looked, for example, at what and who moved in and out of facilities on a daily basis. Often, people in detention centres referred them to compatriots and friends in other facilities whom we then visited.

There were, of course, limitations in this approach and further ethical dilemmas encountered in the field, discussed as the conversation continues in this article. In the case of detention on islands, field research took place in sites that remain inaccessible to most residents of Australia, the United States or Italy (in the cases of Christmas Island, Guam, Saipan, and Lampedusa). Compelled to do this research because of the desire of authorities to hide people in remote detention, we pressed on in spite of ambivalence of researchers, authorities and participants about detention. As Mary Bosworth (2014) argues, while acknowledging this ambivalence, it is important to conduct research inside facilities to show what is happening there. While we agree with Bosworth, we also find that research cannot be premised on 'getting inside', but should involve alternative plans should access be blocked or risky to those detained. Relying solely on 'getting in' risks reliance on masculinist stances which parallel earlier 'muddy boots' principles of geographic field research: the notion that researchers take risks to work in 'risky' areas (Sundberg, 2003). Balancing somewhere between these two positions – that one must have access or that having access is itself problematic – researchers entered the field.

The second project is Author 1's doctoral research on waiting zones (*zones d'attente*) in French airports. The term 'waiting zone' assigns both legal and geographical dimensions. The *Law on the Waiting Zone* established[1] a particular legal regime for 'aliens' (i.e. non-citizens, French: 'étranger') denied entry into French or Schengen territory or claiming asylum at a French border arriving by train, boat or airplane (CESEDA, 2016 articles L221-1 to L224-4). Individuals located in the waiting zone are not governed by ordinary laws, but instead subjected to its particular legal regime. For example, the regular refugee determination procedures do not apply to asylum seekers at the border. Unaccompanied minors are denied due process rights that children in France enjoy (Human Rights Watch, 2014). This parallel – and less protective – legal framework builds on the premise that rejected non-citizens and asylum seekers at the border find themselves at the threshold of, but not having yet entered French sovereign territory. They find themselves, instead, in territorial border zones where the state establishes the distinction between physical and legal entry: physical presence proves insufficient and only lawful admission amounts to entry into the territory (Basaran, 2011). Individuals falling under the purview of the waiting zone framework are detained during the time necessary for them to be returned or, if they are claiming asylum, to determine whether or not their claim is inadmissible or 'manifestly unfounded'. Only those passing this initial screening will be entitled to enter French territory where claims will be examined on their merits.[2] Initially (in the early 1990s), those denied entry were geographically circumscribed by an administrative authority (*le préfet*) to waiting zones, which ran between points of boarding/disembarkation and border checkpoints. These zones could include accommodation located on or nearby the airport, port or train station.

In 2003, the definition of the waiting zone was extended to include any place where an entering non-citizen goes for administrative or medical reasons. As a result, the waiting zone status now accompanies the individual, even kilometres from the point of arrival. For example, individuals turned away at the border at Charles de Gaulle (CDG) airport are deemed outside of France under immigration law when presented to a judge at the Bobigny court – located 17 km from the airport – after four days of detention. Trapped by the waiting zone framework, aliens 'are not expelled by the border, they are forced to *be* the border' (Khosravi, 2010, p. 99). Thus the term 'waiting zone' refers both to a geographical and legal space. On paper, there are 67 waiting zones in France, all located in ports, airports and train stations open to international passengers (Anafé, 2015, p. 13). Most are not significant crossings and therefore used infrequently. Since 2011, ad hoc waiting zones are also created when a group of 10 non-citizens arrives in France outside of a border crossing (e.g. on a beach). In this case, the waiting zone will run between places where authorities have found the non-citizens and the closest border checkpoint.

Author 1's research focused on Charles de Gaulle airport, France's main and Europe's second-largest airport and its waiting zone – which features a built-in detention center (*Zapi 3*) (Human Rights Watch, 2009). As a result of serving more than 60 million passengers every year, CDG's waiting zone detains the greatest number of individuals in France; in 2011, 6,013 individuals were denied entry (Aéroports de Paris [Paris Airports], 2012, p. 7; Anafé, 2013, p. 18) and more than 88% of asylum applications at French borders were made there (OFPRA, 2012, p. 34).

Data were collected from 2013 to 2014 with the goal of understanding how different actors working in and on the waiting zone comprehended this place and their role therein. The project also sought to understand how detainees were treated by different actors working inside the waiting zone.

The final project, carried out by Author 3, explores how nation states were shifting territorial borders using migrant boat interceptions in the Central Mediterranean Sea from 2006 to 2014. During this period, at least 262,128 migrants on 1,681 boats were intercepted attempting to enter Italy and Malta from North Africa (Communities of the European Communities, 2008; Frontex, 2014a, 2014b; Ministero dell'Interno, 2014a). Frontex, the European Union's (EU) coordinating border agency, and the Governments of Italy and Malta have spent at least 67.7 million Euros to conduct regular maritime interdiction operations and prevent seaborne migrants from reaching their shores (Frontex, 2015b; Marina Militare, 2014). Journeys and interceptions in the region are dangerous, leading to at least 3,072 recorded deaths in 2014 (International Organization for Migration, 2014). Those who survive are often diverted, detained or returned, with at least 9592 migrants held in detention in Italy as of 30 December 2014 (Ministero dell'Interno, 2014b). The fate of those intercepted at sea is shrouded in secrecy and statistical gaps. While 64,517 migrants were reported as intercepted in 2013, for example, the Government of Italy only issued a total of 31,270 refusal of entry and illegal presence/expulsion orders that year (Eurostat, 2015a, 2015b). What, then, happened to the majority of migrants intercepted at sea?

The impetus for this doctoral research followed Author 2 and Author 3's discovery and recording of 175 boat loss incidents involving nearly 20,000 migrants since 1990 (Williams & Mountz, 2016).[3] Questions emerged surrounding the activities of state operations designed to intercept migrants, the answers to which were largely unavailable in media reports and scholarship. After following a number of leads, we found a complete listing of Frontex's operations at sea with limited descriptions of characteristics, which prompted more detailed investigation. Subsequent freedom of information requests to the organization resulted in the release of over 600 pages of operational documents (Frontex, 2015a). Over half (or 303) of these pages, however, were covered with redactions, some as long as 16 entire pages, 'in the interest of public security'. The redacted material, like the statistical gaps, described some of the most important information about these operations, such as participating actors, locations of the interceptions and the fate of the migrants involved.

Problems arose regarding which research methods proved appropriate when states attempted to suppress information. These problems were important because authorities often used migrants as justification for securitization and state-building (Bigo, 2002; Walters, 2008). The researcher therefore proceeded by critically digging through the redacted materials and missing statistics to compare and fill gaps in data using unconventional sources and earlier academic literature. A wide variety of data provided hints into the missing pieces and sensitive nature of migrant maritime interdiction operations. They revealed flaws and contradictions in established narratives regarding state and migrant activities in the Central Mediterranean, forcing the researcher to reimagine processes that were rendered explicitly unobservable. This meant trying to find information intended to be secret in the name of public security.

Each of these three projects holds a common desire to study practices and spaces of confinement designed explicitly to contain, immobilize, hide, deter and exclude people on the move. Each happened in a heavily politicized field where authorities are able to hide information. Each involved significant barriers to gathering data, ranging from fortified walls that literally blocked entry (exemplified by the photograph in Figure 1) to more subtle efforts to control information and access. In each case, these barriers shaped the methods used.

Figure 1. Photograph of North West Point Immigration Detention Centre on Christmas Island, overseas Australian territory. Photograph by Alison Mountz.

3. What compelled us to do research in difficult settings with marginalized groups? In other words, why does this matter?

Malkki (1996) demonstrates how humanitarian interventions tend to silence refugees by deliberately disregarding their particular political, historical and cultural contexts in the search for signs of suffering. The international community often constructs refugees as pure victims, devoid of individuality and denied status as credible participants. As a result, their voices remain too often absent from public discourse and public policy-making. Similarly, Myriam Ticktin (2011) argues that people seeking refuge must be constructed as victims in order to be 'morally deserving' subjects of humanitarian assistance. On the one hand, refugees' and asylum seekers' accounts of their own conditions are often disqualified as they are constructed as helpless victims incapable of speaking for themselves. On the other hand, research itself can also reproduce this tendency to construct migrants, refugees and asylum seekers as victims. When not included in the research process, people may become 'objects' of research, distanced through techniques that present them in objectifying and dehumanizing ways, or what Michael Brown (1995) calls 'the ironies of distance': the effect of objectifying and dehumanizing people in an effort to document an empirical phenomenon. Our research projects sought to negotiate and navigate some of these problems. Whenever possible we tried to do research *with* people construed from a distance as vulnerable. Our research was built upon the premise that refugees and migrants' voices are valuable and indispensable in understanding processes of human migration; and that their inclusion involves our geographical proximity.

Author 2 aimed to shed light on processes of externalization – the construction of offshore enforcement and detention measures – where violation of international law and evasion of the commitments of signatories to the 1951 *Convention Relation to the Status of Refugees* had become normalized. In the process of intercepting and detaining offshore, life for migrants became more precarious. With more risk-taking to reach sovereign territory, more 'border deaths' have happened (International Organization for Migration, 2014; Weber & Pickering, 2011). Furthermore, states operate on islands in ways that detach and distance those seeking protection from view of publics as well as journalists, human rights monitors and others who might advocate on their behalf. This distancing propelled forward a research plan fuelled by the desire to counter distance with proximity, hiding with documentation.

Author 1's research aimed to understand the enforcement practices and legal mechanisms that underpin and make Charles de Gaulle airport's waiting zone possible. Indeed, despite civil society organizations' many years of activism, people detained at the airport are still held under a parallel legal framework. Results challenged the idea that a parallel and less protective system of rights based on the distinction between physical and legal entry was normal and necessary in a liberal democracy bound by human rights and refugee conventions at the regional and international levels.

Author 3 found that previous scholarship on boat interceptions around the EU was unable to fully analyse the macrogeographies of migrant exclusion due to lack of information. At the same time, a number of scholars provided key insights into the mechanisms by which states were creating new borderlands at sea (Carling & Hernández-Carretero, 2011; Gammeltoft-Hansen, 2008; Mountz, 2010). Based on clues provided in the Central Mediterranean, it became apparent that a large but unknown number of people travelling by boat were being intercepted, diverted, detained or returned (Andrijasevic, 2006; Bialasiewicz, 2012). States were increasingly creating and invoking extra-legal procedures which resulted in the processing of seaborne migrants en masse and led to effective expulsion from EU territory before arrival ever technically occurred, justified by security and humanitarianism (Betts, 2006; Carling, 2007; Pugh, 2004). Public discourse, however, paid little attention to these issues and reproduced securitizing narratives centered on trafficking, border security and loss of life at sea (Williams & Mountz, 2016). These processes removed the agency of migrants. The research sought to bring attention to the situation in the Central Mediterranean in ways that restored human agency to migration processes.

It is important to carry on with research on boat interceptions, despite data being hidden, to begin forming a coherent empirical picture and bring light to the securitization of migration. Unlike other studies in geography, data analysis using traditional methods, like time-series or spatial analysis, is often not possible with politically sensitive subjects. As discussed in our next section, a mixed methods approach using data from a variety of sources should instead be used to triangulate a coherent empirical picture across scales (Silvey & Lawson, 1999). The development of this approach also supports future work in related, politically sensitive subjects by giving or outlining ways to form an empirical foundation. Research on boat interceptions, for instance, has been advanced by geographers studying the deployment of spatial tactics to manage vulnerable populations (Cresswell, 2006; Hyndman & Mountz, 2008; Samers, 2010). Production, centralization and dissemination of data on politically sensitive subjects bring scientific legitimacy to alternative narratives which challenge existing social hierarchies, fighting the removal of migration from public discourse via securitization (Bigo, 2002; Boswell, 2003).

It is thus important to press on for myriad reasons in spite of – and sometimes precisely because – of missing data and barriers to access. Finding ways forward in these circumstances, however, required creative exploration of research methods, including mixed methods and collection of both formal and informal knowledge about migration and enforcement.

4. How did we do this? An overview of the methods employed and why

Increasingly, geographers are following and forging participatory research models (PAR) as a way of confronting power asymmetries in the field (e.g. Kindon, Pain, & Kesby, 2007). In this model, researchers collaborate in struggles for social justice as an ethical stance with marginalized groups: 'a practice of researching with rather than on participants' (Cahill, Sultana, & Pain, 2007, p. 312). While compelled by this approach, it is sometimes not possible to work in the model of PAR with people detained in confined spaces or intercepted at sea. We sought other ways to work accountably and carefully. We proceeded with the principle of 'do no harm' firmly in mind, prioritizing the importance of not putting people in further danger as a result of research, wherever possible (Mackenzie, McDowell, & Pittaway, 2007).

Research on island detention and in the waiting zone employed archival work and qualitative methods involving human subjects: semi-structured interviews and participant-observation. Interviews were the only way of grasping how different actors (from civil servants to lawyers and NGO volunteers) understood architectures of detention and waiting zone, and their roles within these sites. However, the interview method was employed carefully, in places, times and with participants who would not be rendered more vulnerable as a result of participation – as far as we could understand or imagine. For the waiting zone and island detention projects, most interviews took place outside of detention facilities: detainees themselves were therefore not interviewed, but former detainees were.[4]

We had to remember that authorities can also be vulnerable participants: they are workers whose livelihood and employment status may be jeopardized by disclosures made to researchers. When conducting participant-observation in the physical detention spaces of CDG's waiting zone (i.e. small holding cells in airport's terminals and the detention center) and island detention facilities, it was essential to understand the interactions of public and civil society stakeholders with detainees. As Herbert (2000, p. 552) observes, participant-observation not only understands how people *narrate* their work, but how they *do* their work. There were legal and practical challenges to using these methods.

First, Author 2's research design had to take legal limitations into account. On Christmas Island, employees of the high security North West Point Detention Centre are contractually required not to discuss their jobs outside of work. This necessitated use of participant-observation over interviews, particularly on a very small island where conversations and interactions in a small population and restricted built environment were not easily hidden. Likewise, Author 1 found accessing spaces of confinement in airports (themselves highly strategic and secure locations) difficult since journalists and members of the general public are banned from these spaces. As French law authorizes some NGOs to enter waiting zones, the researcher eventually managed to get access as a volunteer with Human Rights Watch (HRW). The organization had published a report in 2009 on unaccompanied minors and the researcher offered to help update it.

Secondly, we empathized with Martin in Belcher and Martin (2013) who, like us, 'entered a field in which the key actors were polarized by mutual distrust and suspicion [...]' (p. 407). As researchers working in politically charged arenas, we had to manage informants' perceptions of ourselves and keep in mind Hyndman's (2001) observation that the fieldworker is not an outsider, but implicated in the social relations she studies. Doing research with an NGO comes with a host of challenges. Jeffrey (2013), for example, found it difficult to manage NGO expectations in post-Dayton Bosnia when he switched from worker to researcher, due to his privileged access to international organizations. On the waiting zone project, many years of antagonism between authorities and activists rendered the concurrent use of interviews and participant-observation impossible. Indeed, Makaremi had warned the researcher of the dangers of being associated with an NGO: her volunteer position with the activist organization Anafé had precluded interviews with government officials and border police (C. Makaremi, personal communication, 2013). The researcher therefore attempted to interview as many officials as possible before visiting the airport's spaces of confinement with HRW.

At the airport, Author 1 met a policeman at a terminal's reception desk who confided the torments that his job entailed: some of his colleagues were racists and wanted to close borders. He told the researcher he had a lot more to say and wanted to meet again, but outside of his workplace. However, he was visibly tense and scared. He explained that he would call from a public phone and use a pseudonym to schedule a meeting. But he did not call, and was therefore never interviewed. The researcher saw him again on another visit. While she was certain he possessed invaluable information, she did not press him to speak further. As Malkki (1995) observes, researchers have to demonstrate their trustworthiness by not prying where they are not wanted. We must accept that it is not possible to know everything and realize the value of the partial information delivered by informants (Malkki, 1995).

Similarly, the field research on island detention involved many informal exchanges where potential participants in research refused formal interviews, yet shared valuable insights in more informal settings. These settings were usually outside of the immediate workplace, in island cafes or restaurants, for example. The participation of authorities, whether through formal interviews or informal exchanges, supported the notion that the state does not function as a monolith. Indeed, workers in waiting zones and detention centres express different viewpoints and emotions about these sites and their work, and these disagreements may make them vulnerable as individuals, employees and participants in research.

Participation among visitor groups, advocates and activists also rendered them vulnerable in different ways. As the neoliberal state has retrenched and withdrawn social support to asylum seekers and others, visitors and advocates have taken up the task of providing material and other kinds of support, such as mental health advocacy for people in detention. This has left many people in small communities exhausted, cynical and vicariously or contagiously traumatized, as Kate Coddington (2016) argues.

For studying interceptions at sea, a mixed methodology was constructed using the extended case method (Burawoy, 1998). Missing data were the major methodological challenge and the justification for this approach. Data gaps are a frequent and recurring problem in research on human migration (Bijak, 2011, pp. 16, 17; Jandl, 2004; Massey & Zenteno, 1999). Traditional issues associated with these gaps, such as inconsistent definitions, measurements and frequencies or geographies of measurement, are compounded by the political

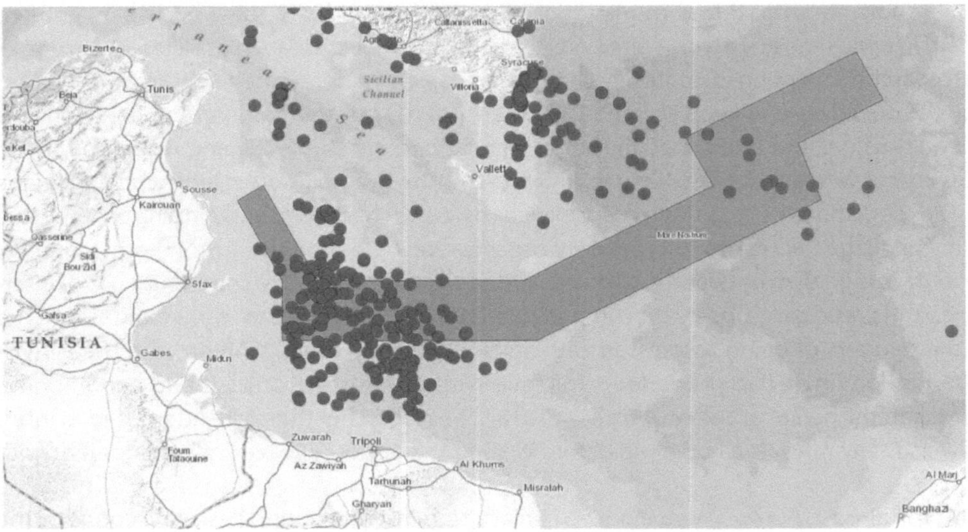

Figure 2. Map of reported boat interceptions during operation Mare Nostrum, 2014 Shaded area at sea: operational area for Mare Nostrum Circles: reported interception. Source: Ministero dell'Interno (2014a).

sensitivity of boat interceptions (Ellis & Wright, 1998, pp. 129–131; Hoffman & Lawrence, 1996, pp. 35–42; Massey & Capoferro, 2008, pp. 259–265). This sensitivity manifests in power relations that alter data definition, collection and storage by influencing research objectives and uses (Burawoy, 1998, p. 23; Ley, 2003, pp. 432–435; Silvey & Lawson, 1999, pp. 124–126). In the case of interceptions at sea, authorities may perceive a need to falsely measure, collect, report or even destroy data. One curious example of this phenomenon conjoins the projects of Authors 2 and 3. While conducting research in Lampedusa, Author 2 found that the Government of Italy had reported no boat arrivals in the region during 2010, despite contrary reports by locals. In 2014, we were also informed by Frontex that boat interception data for Italy in 2010 were missing despite the existence of similar data during previous years of the same operation (Frontex, 2014a). We acquired operational documents that not only confirmed the existence of an operation in the region during 2010, but also demonstrated that Frontex actively collected these 'missing data' in regular incident reports (Frontex, 2015a). Elsewhere, scholars argue that EU member states manipulated nationality statistics in order to 'repatriate' individuals to third countries with which they have readmissions agreements (Andrijasevic, 2006; Carling, 2007).

Author 3 found that analysis of partial information became necessary to triangulate important findings. Conclusions supported by this analysis served to extend theory through refutation as well as identify new data sources. These new sources, in turn, were reflexively incorporated and compared to previous findings to update conclusions and further extend the project until a substantial body of data was acquired. This led to the use of a diversity of data sources recording different information, such as migration enforcement, asylum applications and outcomes, international treaties, operational documents, academic literature, media reports and even diplomatic cables. Data often had to be constructed in unconventional ways. For example, as shown in Figure 2, the researcher converted a map provided by the Government of Italy's Ministry of the Interior showing the location of all boat interceptions around Operation *Mare Nostrum* in 2014 into numeric counts of interceptions by region (Ministero dell'Interno, 2014a). The number of people intercepted but never recorded as 'arrived' in Italy was constructed by using an accounting method with multiple sources of enforcement reports. Finally, the researcher made a database storing the characteristics of all redactions made to the Frontex documents he received. After extracting the text from these documents using text recognition software, missing data analysis was used to infer the topics of the information contained within the redactions. By piecing together these diverse sources of information, a coherent empirical picture of boat interceptions emerged. While each separate source lacked complete validity or reliability, the collected data, together, were sufficient to test and extend theory on spatial tactics related to interception.

5. What subsequent problems and dilemmas did we face once research began?

We each faced myriad issues while collecting data, beginning with barriers to access. We describe these barriers here before explaining how we resolved them in section VI. Although immigration law and visitor protocols at detention facilities allow for visitor access, in practice this access varies greatly according to the authorities in charge on any given day. The researcher who enters into confined spaces does so amid complex, if well established, power relations. Those confined and their visitors are subjected to arbitrary and diverse rules, varying greatly from one authority to the next. The very same official might display distinct attitudes as time unfolds. For example, during the first visit to Charles de Gaulle airport, Author 1 witnessed a border officer's transformation: within an hour he went from being welcoming and eager to please to quiet and aggressive, reaching a point where he stopped answering questions. At the end of the visit, he explained the reason for his anger: his colleagues had not come to relieve him of his duties. Having started his shift at 5 am, by the time of the lunchtime visit he was starving. In a place where people must rely on authorities for everything – in some terminals, detainees have to knock on the cell door from inside and hope that police will have time or inclination to allow them to use the bathroom – every change in mood or well-being affects both detainee's and researcher's access to information.

Similarly, Belcher and Martin (2013) explain that policies of access to US detention centers 'changed randomly and without notice, or were interpreted differently by detention center staff members' (p. 408). One weekend in February, the researcher experienced first-hand at Charles de Gaulle what Martin had observed in the US. The police officer accompanying her attempted to silence both researcher and detainees by affirming that she could not discuss any 'procedures' with the individuals confined in terminal 2E, claiming her mandate was only to verify detention conditions. This precluded any conversation about rights or administrative or judicial procedures pertaining to the detainees' status. When challenged with the observation that these requirements were nowhere to be found in law, the policeman replied that they were specified in an internal policy note which the researcher could not see as she was not a civil servant. The visit posed a significant ethical dilemma: the officer, standing less than one meter apart from her and detainees, listened closely to the conversations. As will be explained below, this illustrates the fact that researchers must try to adapt to situations quickly but ultimately prioritize research participants' well-being over data completeness.

[redacted]

FRONTEX

[redacted]

Figure 3. Sample of a redacted document from Frontex (2015). Source: Frontex (2015a).

In research on interceptions at sea, the barriers to accessing missing data became personal when Author 3 received hundreds of pages of redacted documents from Frontex, exemplified by the excerpt shown in Figure 3. These documents were accompanied by two letters from the head of the Legal Affairs Unit which claimed that the reproduction of information contained therein may be restricted due to international copyright law and provided 15 days to make an appeal to re-assess the redactions (Frontex, 2015c). The researcher found this appeal difficult to make, since it was impossible to know with certainty what information the redactions had removed or whether they were valid exemptions in the interest of public security. The redactions were additionally confusing because multiple redactors had taken part and used inconsistent redaction techniques, meaning that information which was likely meant to be redacted sometimes remained present. Contact information for key personnel or offices, for instance, was available in some documents though redacted in most. More surprisingly, an entire, unrelated mission report from Frontex's Return Operations sector was accidently contained within the operational documents (Frontex, 2015a). It seemed likely that the discovery of this leak would have repercussions, and he therefore debated how to discuss it without harming those involved at Frontex. Who could he contact to discuss these issues? What could he do next? How should he carry on with research, given that authorities were both determined to and ineffective at hiding what he wanted to know?

Ethical dilemmas also arose when we learned about poor conditions, policy violations and violence transpiring in confined spaces. Like the hidden data inadvertently released, we wondered how to share this information and with whom, who was at risk of further violence and who was at risk of losing employment as a result. There were never easy answers. For instance, at Charles de Gaulle during another visit in February, the researcher was confronted firsthand with instances of mistreatment. While she was conversing with Anafé's volunteers at the detention center (the NGO holds a legal aid clinic inside the facility twice a week), a group of six women from Central America burst into the office showing signs of great distress. They recounted mistreatment by police during the attempt to force them on the plane and showed multiple contusions on their bodies. One of the women had a certificate from the facility's doctor documenting the wounds. Anafé's employees told the volunteers that the women could file a complaint if they so wished. Once they agreed, the researcher found herself in an uncomfortable position: none of the volunteers spoke Spanish and she was therefore the only one able to collect their testimonies. However, Human Rights Watch's (HRW) mandate (which she had to embrace, as a HRW volunteer), unlike that of Anafé, precluded intervention in individual cases. Intervention therefore meant overstepping bounds. Moreover, she was aware that her involvement could result in sour relations between authorities and HRW, not to mention jeopardizing her own access to the facility. Our next section details how she proceeded.

Similarly, researchers gathering data on island detention grappled with damaging information shared in informal settings by employees of facilities, including violence enacted within facilities. They knew that sharing information publicly could endanger the livelihood of workers, render people confined more precarious, endanger access of advocates to facilities, and that sharing information publicly could also potentially violate conditions of research procedures approved by university ethics boards.

In addition to ethical challenges, our research triggered many different emotions: anger, frustration, anxiety, sadness and vicarious trauma. As Bondi (2005) demonstrates, this is a normal part of research: researchers (whether using quantitative or qualitative methods) always enter relationships when collecting or generating data: they are drawn into 'relationships with texts', 'numerical data-sets' or 'with the interfaces through which such data are extracted or analysed' (p. 236). Unsurprisingly therefore, researching information that authorities try to suppress can be emotionally challenging.

Anthropologist Ruth Behar (1996) argues that researchers are 'vulnerable observers', and other scholars have written about this process as a kind of vicarious trauma (Coles, Astbury, Dartnall, & Limjerwala, 2014): 'the transformation of the therapist's or helper's inner experience as a result of empathetic engagement with survivor clients and their trauma material'

(Pearlman & Saakvitne, 1995; cited in Coles et al., 2014, p. 97). While the term has been used primarily to describe experiences of professional counsellors working with rape victims, experts explain that anyone can be vicariously traumatized after having had extended contact with trauma victims (Wasco & Campbell, 2002). All researchers on these projects underwent psychological impacts that mirror physical manifestations of stress on people in detention (albeit in significantly milder form), including anxiety and depression. These feelings were compounded by a sense of isolation. It is difficult to share them with people not working on these subjects, who may feel that our unease is exaggerated, since we did not experience the traumatic experience first-hand but just learned about it. On the boat interception project, the investigator became aware of more complex, problematic imaginations of what has likely transpired as established narratives on the activities and outcomes of operations are refuted. More often than not, this entails large-scale violations of human rights. What was nearly as crushing as this human cost was the researcher's perception of apathy in the discourse for the worth of the lives involved. As Coles et al. (2014) observe in their research on sexual violence, researchers are more exposed to vicarious trauma than clinicians or counsellors precisely because of their inability to help people who survive abuse.

Moreover, researchers repeatedly exposed to traumatic events may experience 'saturation' (Coles et al., 2014). Justified by security and humanitarianism, for example, the stories of seaborne individuals are normalized and obscured; they are compiled into the final resting places known only as missing data. Author 3 thus felt it critical to not repeat the objectification of those involved in boat interceptions through the production and collection of new data. At the same time, it became impossible to consider himself an unbiased observer. An ethical dilemma occurred, then, on how to conduct research related to boat interceptions without reproducing its social hierarchies and exclusions through detachment. In this way, the researcher determined to use his research as a form of resistance (Sharp, Routledge, Philo, & Paddison, 2000).

As Bondi (2005) and Rose (1997) note, interviews and participant-observation are methods that rely heavily upon interpersonal encounters; the very interaction between researcher and informant creates data. This immersive, proximate mode of data collection leaves the researcher particularly permeable to emotions experienced by informants (Coddington, 2016). Moreover, people conducting research with marginalized groups such as asylum seekers also often experience forms of guilt and paralysis (Mountz, Miyares, Wright, & Bailey, 2003). In our experiences, we felt guilt about collecting stories or committed to not ask people to tell stories that might have forced them to revisit trauma. Asylum processes are designed to reward traumatic histories of violence, requiring applicants to tell stories in particular ways. Over time, their stories become commodified in ways that we wished not to reproduce through the research process. In the case of participant-observation and interviews conducted, 'truth' was not the objective of inquiry, given that all such encounters are staged performances of one kind or another (Katz, 1992; Mountz et al., 2003; Rose, 1997). Instead, we aimed to encourage participants to direct the conversation, a technique of feminist research methods (Anderson & Jack, 1991).

For all of these reasons, we concur with Hyndman's (2001) assessment that field research changes the researcher as much as the researcher changes the field. It becomes ethically impossible to leave behind or ignore these experiences. There is no such thing as detached observer, but rather researchers hold dual status as simultaneously vulnerable and powerful.

6. How did we move forward in spite of these problems?

Doing research in politicized fields is particularly challenging, as there is no blanket solution which works in all settings. Researchers navigate particular issues as they arise, often requiring the researcher to make important decisions in haste. These decisions may subject people to retribution or other vulnerability, especially if their details are published. We felt it critical, then, to consider the political ramifications of withholding information (e.g. regarding violence) for publication on the people involved in our studies. What happened next after issues arose was often contextual to each project, and clear solutions were uncommon.

As far as Author 1 was concerned, she attempted to bypass the police officer's illegitimate restrictions on interaction by first engaging with English- and Spanish-speaking detainees in their language. When it became apparent that the police understood neither English nor Spanish, the researcher could still gather some valuable information, when she heard about an allegation of mistreatment at the hands of the police of a woman named Maria[5] for the second time. This incident probably explained why the authority did not want her to speak openly with detainees.

However, this strategy was limited by the common word origins of French, Spanish and English. The policeman put an abrupt end to the conversation after hearing the word 'court', which is the same in French and Spanish (*tribunal*). It was particularly heartbreaking for the researcher not to be able to explain asylum procedures to an exhausted and obviously traumatized Syrian asylum-seeker (who said repeatedly, 'Please, don't attack me'). The officer told the researcher it was forbidden to discuss 'political asylum'. When it was time to speak to Francophone detainees, the researcher asked the only question she could think of that would not endanger them: Had they eaten? The answer to this question was evident as garbage and picnic leftovers were scattered all over the room. Fearing that detainees could be the targets of retaliation if they complained, she chose to ask the minimum amount of information possible. Ultimately, it is the researcher's ethical duty to decide when to stop asking questions and leave, however frustrating the situation.

Regarding the instances of violence, being an 'unbiased observer' proved impossible, and Author 1 felt compelled to collect testimonies, even though the repercussions of her involvement were impossible to predict. Her participation (as her name appeared on the complaint's file) affected how the police received her during subsequent visits. While invited for coffee in the police staff room in the afternoon of that same day (before the events unfolded), she was met coldly and even with outward hostility afterwards, as if she had betrayed the trust that police had placed in her. She immediately reported the incident to HRW and luckily it did not affect her relationship with the organization. Nor did it sour relations between authorities and HRW. As for the six women who filed a complaint for allegation of police violence, two were returned shortly afterwards, while four were sentenced to a few months in prison by a criminal court for remaining illegally on French territory. Their complaints never went through; the state prosecutor never contacted them, and thus the police were not held accountable for their actions.

While the limitations of participation in research were frustrating, Author 3 found designing and implementing a methodology to overcome institutional barriers was the best way forward. At first glance, the desire to explain the activities and outcomes of boat interceptions on a large scale proved daunting. There were no established methodologies and data seemingly non-existent. Previous scholarship had adopted an eclectic set of approaches, and it was impossible to say what exactly he would be able to find. These issues limited the ability to construct a study in the 'quantitative' ways common in human geography literature. This also made it difficult to compete for academic funding, since having a firm research plan and methodology at the outset of the project was unrealistic. Given the need to be flexible during research in order to handle its ethical challenges, this rigidity of funding itself may be unethical.

A pragmatic solution to these problems was to design and build a new methodology. The researcher began by checking the initial academic sources that had prompted the research and reviewing every data source used. This process generated a number of obscure but relevant information sources and potential methods. For example, Author 3 learned from Author 2 and Weber and Pickering (2011) that Marg Hutton (2015) had collected a large body of documents on boat interceptions in Australia. Investigating Hutton's project demonstrated the use of freedom of information requests in collecting previously unpublished data, which were then used in the case of Frontex. By using these initial sources and probing additional methods, it turned out that a variety of data existed. A mixed methodology was selected to make use of the diversity of evidence and to focus on generating and testing theories of spatial tactics with respect to boat interceptions. Although further data gaps emerged since that time, each instance of missing data proves important in uncovering how data are produced and the power relations underlying them.

7. What did or can we do with this information once collected, and what are some of the ethical issues associated with data analysis and dissemination?

If research is to be a form of resistance, then it must be used to challenge dominant narratives around 'vulnerable' populations, including those which (re)produce violence through the creation and enforcement of social hierarchies. Like others (e.g. Cahill et al., 2007), we explored how the tools of research can be put to use to enhance society and advance social movements. In our collective experience, this prerogative stems from researchers' desire to participate in research and related political discourse as well as the burden of knowledge generated therefrom. In the case of boat interceptions in the Mediterranean, we seek to put methodological tools to use in a helpful way. This means, first, to produce empirical knowledge on the missing data and, second, to convey this knowledge to a wider audience. The primary method of dissemination will be to publish data produced and used in the study online in a number of formats. Like Hutton's sievx.com, setting up a website to store and share data is a low-cost way to allow a wider audience to make use of empirical knowledge produced. This measure also satisfies positivist values of open knowledge and reproducibility. Publishing empirical information on missing data lends scientific legitimacy to alternative narratives about seaborne migrants. At the very least, this information adds context and awareness to issues around ongoing boat interception operations. The website will also introduce theory and analysis resulting from the data, which will position alternative narratives in the discourse.

Researchers on the island detention project also sought to enhance public knowledge about detention and offshore enforcement. Mapping and timelines created from the project contributed visual dissemination of information to public arenas, adding to growing social movements to document, memorialize, and map migrant crossings, detention, death and deportation. Some recent examples of this included the Australian Border Deaths Database and Deaths at the Border Database in the European Union (Border Crossing Observatory, 2016; Last, 2016). Researchers also participated in public discussion and debate, through radio and televisions shows and interviews, podcasts and blog posts.

Publication holds risks for precarious populations by influencing the behavior of those involved with boat interceptions. Part of Frontex's justification for redactions in the interest of public security, for example, was that public release of operational details or outcomes could hinder future operations. This is possible because migrants and smugglers will change tactics to prevent interception by authorities. In situations where migrants face constrained resources, these changes will affect their livelihood. In 2014, for instance, Frontex claimed that Italy's Operation *Mare Nostrum* was drawing more migrants by being closer to Libyan shores, since migrants only needed to make it to Italian vessels to claim asylum (Frontex, 2014b).[6] By withdrawing patrols vessels from these areas when the operation ended in 2014, it is likely that a large number of people on boats were left stranded. Authorities or migrants may also be harmed by exposing illegal activities, regardless of intentionality. Legal violations made by migrants may be used to justify further externalization and securitization, generating a vicious cycle. Publication may even affect future researchers when actors decide to change the data construction process. Developing a methodology to examine redactions, for instance, may influence organizations like Frontex to change the way in which they are made to prevent future inference. All of these cases highlight risks to multiple groups resulting from publication of data and findings. A balance, then, should be found between the benefits of dissemination and the potential harm, to the extent that this harm can be anticipated.

8. Conclusions

In this article, we explored some of the barriers to and challenges associated with research with marginalized groups in difficult-to-reach locations, whether spaces where people are forcibly confined or wide-open spaces such as the Mediterranean Sea. We explained which methods we chose to overcome institutional barriers to access information about vulnerable groups while taking their welfare into account. Before, during and after conducting these projects, we found no easy answers, but build here on the premise that deeper discussion helps and must triumph over silence and isolation.

By geographically confining people on the move, states create both a geographical and emotional distance between citizens and non-citizens. They purposely obscure views of the 'other', which tends to reproduce and reinforce myths about migrants. As Agier (2008) observes, by isolating people states convey the message that there is something wrong with them: they must be isolated for a reason. The ingrained idea that citizens of any given country are different from non-citizens constitutes one of the major obstacles to non-citizens' enjoyment of human rights, according to Francois Crépeau, United Nations Special Rapporteur on the human rights of migrants (personal communication, 1 July 2013). It is our hope that dissemination of research findings will challenge this imagined landscape by reducing the gap between 'us' and 'them'.

We explored and simultaneously challenged meanings of vulnerability, by discussing migrants and asylum seekers who may well find themselves in precarious situations as they flee violence, but who nonetheless exercise agency and make decisions along the way. These include decisions about whether and how to migrate, whether and when to claim asylum, whether and when to report abuses and poor conditions, whether to participate in research and under what conditions.

We also explored the positionality of researchers, examining the various webs of vulnerability and power that they too possess and exercise during data collection of all kinds. Researchers working in politicized fields do not have the option of shying away from difficult issues and encounters. They must wade through thorny entanglements of vulnerability and power that transpire on perpetually shifting terrain. Dynamics at work one day may shift the next; speaking and writing about truths, rules or solutions are therefore difficult. One productive way forward that we have found is to share discussions, emotions, findings, difficulties and decisions. Discussing and reflecting upon the emotional toll of research on and in politicized settings proves helpful in itself as there is more power as a collective. This is why we find a need for this conversation to continue beyond the pages of this article. These conversations must also involve other people in the field: asylum seekers, migrants, detainees, NGO workers, activists, service providers, authorities and policy-makers.

It is too easy to become thwarted, blocked from access to information in politicized fields of knowledge; too often, paralysis threatens. But we do not like the alternative in our respective fields. What if no one actually gathers this information? What if voices of people themselves – whether migrants experiencing enforcement practices or authorities carrying them out – are not included? These exclusions would add to the silences of research and analytical understandings of these phenomena. We are challenged and changed by this work, but also politicized to ask new questions and keep doing research.

Notes

1. The law was initially adopted in July 1992. Since then, its scope has been significantly extended.
2. Since the law of 29 July 2015, asylum seekers may also be admitted to the territory when deemed particularly vulnerable by the refugee agency. When not considered as such, unaccompanied minors seeking asylum still have to pass the 'manifestly unfounded' test, except for a very few cases.
3. This research is related to, but not one of the three projects discussed in this paper.
4. Some managers were interviewed inside facilities. Unlike detainees, managers had power to control conditions in order to speak privately on the job (e.g. in a private office).
5. All references to research participants in this paper are pseudonyms.
6. This claim is misleading, since there are few legal protections for asylum seekers on the high seas or in the seas of a third country. Although Italian authorities allow for asylum claims for 'rescued' migrants on the high seas, they generally do not recognize a 'right to seek asylum'.

Acknowledgements

We deeply appreciate the time, effort, work, and trust of everyone who helped us with our research in countless ways: participants, collaborating researchers, and people who shared their expertise on and in the field.

Disclosure statement

No potential conflict of interest was reported by the authors.

Funding

This work was supported by the National Science Foundation [grant number 0847133].

References

Aéroports de Paris [Paris Airports], (2012). *Aéroport Paris-Charles de Gaulle: Guide du Client 2012* [Paris-Charles de Gaulle airport: 2012 Customer Guide]. Aéroports de Paris: Roissy Charles de Gaulle.

Agier, M. (2008). *Gérer les indésirables: des camps de réfugiés au gouvernement humanitaire* [Managing the undesirable: From refugee camps to humanitarian government]. Paris: Flammarion.

Anafé. (2013, January). *Guide théorique et pratique: la procédure en zone d'attente* [Theoretical and practical guide: Procedure in the waiting zone]. Retrieved from http://www.anafe.org/spip.php?article245

Anafé. (2015, November). *Des zones d'atteintes aux droits. Rapport d'observations dans les zones d'attente et rapport d'activité*. [Zones of right infringement. Report of observations in waiting zones and activity report]. Retrieved from http://www.anafe.org/spip.php?article317

Anderson, K., & Jack, D. C. (1991). Learning to listen: Interview techniques and analyses. In S. B. Gluck & D. Patai (Eds.), *Women's Words: The feminist practice of oral history* (pp. 11–26). New York, NY: Routledge.

Andrijasevic, R. (2006). Lampedusa in focus: Migrants caught between the Libyan desert and the deep sea. *Feminist Review, 82*, 120–125.

Barnett, M., & Duvall, R. (2005). Power in international politics. *International Organization, 59*, 39–75.

Basaran, T. (2011). *Security, law and borders: At the limits of liberties*. London: Routledge.

Bauder, H. (2013). *Why we should use the term illegalized immigrant* (RCIS Research Brief No. 2013/1). Toronto: Ryerson Centre for Immigration and Settlement.

Behar, R. (1996). *The vulnerable observer: Anthropology that breaks your heart*. Boston, MA: Beacon Press.

Belcher, O., & Martin, L. (2013). Ethnographies of closed doors: Conceptualising openness and closure of US immigration and military institutions. *Area, 45*, 403–410.

Betts, A. (2006). Towards a Mediterranean solution? Implications for the region of origin. *International Journal of Refugee Law, 18*, 652–676.

Bialasiewicz, L. (2012). Off-shoring and Out-sourcing the Borders of Europe: Libya and EU Border Work in the Mediterranean. *Geopolitics, 17*, 843–866.

Bigo, D. (2002). Security and immigration: Towards a critique of the governmentality of unease. *Alternatives: Global, Local, Political, 27*, 63–92.

Bijak, J. (2011). *Forecasting international migration in Europe: A Bayesian view*. London: Springer.

Bondi, L. (2005). The place of emotions in research: From partitioning emotion and reason to the emotional dynamics of research relationships. In J. Davidson, L. Bondi, & M. Smith (Eds.), *Emotional geographies* (pp. 231–246). Aldershot: Ashgate.

Border Crossing Observatory. (2016). *Australian border deaths database*. Retrieved from http://artsonline.monash.edu.au/thebordercrossingobservatory/publications/australian-border-deaths-database/

Boswell, C. (2003). The External Dimension of EU immigration and asylum policy. *International Affairs, 79*, 619–638.

Bosworth, M. (2014). *Inside immigration detention*. Oxford: Oxford University Press.

Brown, M. (1995). Ironies of distance: An ongoing critique of the geography of AIDS. *Environment and Planning D: Society and Space, 13*, 159–183.

Burawoy, M. (1998). The extended case method. *Sociological Theory, 16*, 4–33.

Cahill, K., Sultana, F., & Pain, R. (2007). Participatory ethics: Politics, practices, institutions. *ACME, 6*, 304–318.

Carling, J. (2007). Control and migrant fatalities at the Spanish–African borders. *International Migration Review, 41*, 316–343.

Carling, J., & Hernández-Carretero, M. (2011). Protecting Europe and protecting migrant? Strategies for managing unauthorised migration from Africa. *The British Journal of Politics and International Relations, 13*, 42–58.

CESEDA. (2016). Retrieved March 4, 2016, from http://www.legifrance.gouv.fr/affichCode.do?cidTexte=LEGITEXT000006070158

Coddington, K. (2016). Contagious trauma: Reframing the spatial mobility of trauma within advocacy work. *Emotion, Space and Society.* http://dx.doi.org/10.1016/j.emospa.2016.02.002

Coles, J., Astbury, J., Dartnall, E., & Limjerwala, S. (2014). A qualitative exploration of researcher trauma and researchers' responses to investigating sexual violence. *Violence Against Women, 20*, 95–117.

Communities of the European Communities. (2008). *Report on the evaluation and future development of the FRONTEX Agency (COM (2008) 67): Statistical data.* Retrieved from http://www.statewatch.org/news/2008/feb/eu-com-frontex-stats.pdf

Cresswell, T. (2006). *On the move.* New York, NY: Routledge.

Ellis, M., & Wright, R. (1998). When immigrants are not migrants: Counting arrivals of the foreign born using the US census. *International Migration Review, 32*, 127–144.

European Agency for the Management of Operational Cooperation at the External Borders (Frontex). (2014a). *Re: Total migrants and boats detected, intercepted, and turned back during JO HERMES, 2007; 2009–2014. Freedom of information request made through Ask the EU.* Retrieved from http://www.asktheeu.org/en/request/re_total_migrants_and_boats_dete#outgoing-3141

European Agency for the Management of Operational Cooperation at the External Borders (Frontex). (2014b). Concept of reinforced joint operation tackling the migratory flows towards Italy: JO EPN-Triton. Retrieved from http://www.avvenire.it/Cronaca/Documents/JOU%20Concept%20on%20EPN-TRITON%20(2).PDF

European Agency for the Management of Operational Cooperation at the External Borders (Frontex). (2015a). Re: Statistical data documents for JO Hermes and Triton, 2007–2014. Freedom of information request made through ask the EU. Retrieved from http://www.asktheeu.org/en/request/re_statistical_data_documents_fo#incoming-6114

European Agency for the Management of Operational Cooperation at the External Borders of the Member States of the European Union (Frontex). (2015b). *Archive of operations.* Retrieved from http://frontex.europa.eu/operations/archive-of-operations/

European Agency for the Management of Operational Cooperation at the External Borders (Frontex). (2015c). *Request for access to Frontex documents.* Retrieved from http://www.asktheeu.org/en/request/1674/response/6081/attach/2/2015%2001%2022%20Hermes%202007%202014.pdf

Eurostat. (2015a). *Non-EU citizens refused of entry at EU-28 external borders.* Retrieved from http://eurostat.ec.europa.eu/nui/show.do?dataset=migr_eirfs&lang=en

Eurostat. (2015b). *Non-EU citizens found to be illegally present in the EU-28 and EFTA.* Retrieved from http://eurostat.ec.europa.eu/nui/show.do?dataset=migr_eipre&lang=en

Gammeltoft-Hansen, T. (2008). The refugee, the sovereign and the sea: EU interdiction policies in the Mediterranean (Danish Institute for International Studies working paper 2008:6). Retrieved from http://www.econstor.eu/bitstream/10419/44650/1/560120990.pdf

Herbert, S. (2000). For ethnography. *Progress in Human Geography, 24*, 550–568.

Hoffman, E., & Lawrence, S. (1996). *Statistics on international labor migration: A review of sources and methodological issues.* Geneva: International Labor Organization.

Human Rights Watch. (2009, October 29). *Lost in transit: Insufficient protection for unaccompanied migrant children at Roissy Charles de Gaulle Airport.* Retrieved from http://www.hrw.org/reports/2009/10/29/lost-transit-0

Human Rights Watch. (2014, April 8). *France: Unaccompanied children detained at borders: 500 a year held in transit zones.* Retrieved from http://www.hrw.org/news/2014/04/08/france-unaccompanied-children-detained-borders

Hutton, M. (2015). *SIEVX.com.* Retrieved from http://sievx.com/

Hyndman, J. (2001). The Field as here and now, not there and then. *Geographical Review, 91*, 262–272.

Hyndman, J., & Mountz, A. (2008). Another brick in the wall? 'Neo-refoulment' and the externalization of asylum in Australia of Europe. *Government and Opposition, 43*, 249–269.

International Organization for Migration. (2014). *Fatal journeys: Tracking lives lost during migration.* Retrieved from http://www.iom.int/files/live/sites/iom/files/pbn/docs/Fatal-Journeys-Tracking-Lives-Lost-during-Migration-2014.pdf

Jandl, M. (2004). The estimation of illegal migration in Europe. *Migration Studies, 41*, 141–155.

Jeffrey, A. (2013). *The improvised state: Sovereignty, performance and agency in Dayton Bosnia.* Oxford: Wiley-Blackwell.

Katz, C. (1992). All the world is staged: Intellectuals and the projects of ethnography. *Environment and Planning D: Society and Space, 10*, 495–510.

Khosravi, S. (2010). *'Illegal' traveller: An auto-ethnography of borders*. Houndmills, Basingstoke: Palgrave Macmillan.

Kindon, S., Pain, R., & Kesby, M. (Eds.). (2007). *Connecting people, participation and place: Participatory action research and methods*. London: Routledge.

Last, T. (2016). *Deaths at the borders: Database for the Southern EU*. Retrieved from www.borderdeaths.org/?page_id=11

Ley, D. (2003). Seeking homo economicus: The Canadian state and the strange story of the business immigration program. *Annals of the Association of American Geographers, 93*, 426–441.

Mackenzie, C., McDowell, C., & Pittaway, E. (2007). Beyond 'do no harm': The challenge of constructing ethical relationships in refugee communities. *Journal of Refugee Studies, 20*, 299–319.

Malkki, L. H. (1995). *Purity and exile: Violence, memory, and national cosmology among Hutu refugees in Tanzania*. Chicago, IL: University of Chicago Press.

Malkki, L. H. (1996). Speechless emissaries: Refugees, humanitarianism, and dehistoricization. *Cultural Anthropology, 11*, 377–404.

Massey, D., & Capoferro, C. (2008). Measuring undocumented migration. In A. Portes & J. Dewind (Eds.), Chapter 9 *Rethinking migration: New theoretical and empirical perspectives* (pp. 257–286). New York, NY: Berghan Books.

Massey, D., & Zenteno, R. (1999). The dynamics of mass migration. *Proceedings of the National Academy of Sciences of the United States of America, 96*, 5328–5335.

Marina Militare. (2014). *Mare Nostrum operation*. Retrieved from http://www.marina.difesa.it/EN/operations/Pagine/MareNostrum.aspx

Ministero dell'Interno [Ministry of the Interior]. (2014a). *Fenomeno Migratorio*. Retrieved from http://www.interno.gov.it/sites/default/files/dati_fenomeno_migratorio.pdf&usg=ALkJrhg8sC7_4ZuZa5olMNHQbYvRkD_Wkg

Ministero dell'Interno [Ministry of the Interior]. (2014b). *Presenze dei migranti nelle strutture di accoglienza in Italia*. Retrieved from http://www.interno.gov.it/it/sala-stampa/dati-e-statistiche/presenze-dei-migranti-nelle-strutture-accoglienza-italia

Mountz, A., Miyares, I., Wright, R., & Bailey, A. (2003). Methodologically becoming: Power, knowledge and team research. *Gender Place and Culture, 10*, 29–46.

Mountz, A. (2010). *Seeking asylum: Human smuggling and bureaucracy at the border*. Minneapolis: University of Minnesota Press.

OFPRA. (2012, April 11). *Rapport d'activité 2011* [Annual report 2011]. Retrievedf from http://www.ofpra.gouv.fr/documents/OfpraRA2011.pdf

Pearlman, L., & Saakvitne, K. (Eds.). (1995). *Trauma and the therapist: Countertransference and vicarious traumatization in psychotherapy with incest survivors*. New York, NY: W. W. Norton.

Pugh, M. (2004). Drowning not waving: Boat people and humanitarianism at sea. *Journal of Refugee Studies, 17*, 50–69.

Rose, G. (1997). Situating knowledges: Positionality, reflexivities and other tactics. *Progress in Human Geography, 21*, 305–320.

Samers, M. (2010). *Migration*. New York, NY: Routledge.

Silvey, R., & Lawson, S. (1999). Placing the migrant. *Annals of the Association of American Geographers, 89*, 121–132.

Sharp, J., Routledge, P., Philo, C., & Paddison, R. (Eds.). (2000). *Entanglements of power: Geographies of domination/resistance*. New York, NY: Routledge.

Sundberg, J. (2003). Masculinist epistemologies and the politics of fieldwork in Latin Americanist geography. *Professional Geographer, 55*, 181–191.

Ticktin, M. (2011). *Casualties of care: Immigration and the politics of humantarianism in France*. Berkeley: University of California Press.

Walters, W. (2008). Bordering the sea: Shipping industries and the policing of stowaways. *Borderlands, 7*(3), 1–25.

Wasco, S., & Campbell, R. (2002). Emotional reactions of rape victim advocates: A multiple case study of anger and fear. *Psychology of Women Quarterly, 26*, 120–130.

Weber, L., & Pickering, S. (2011). *Globalization and borders: Deaths at the global frontier*. London: Palgrave Macmillan.

Williams, K., & Mountz, A. (2016). Remote control: Analyzing the relationship between externalization and migrant deaths and boat losses. In R. Zaiotti (Ed.), *Externalizing migration management: Europe, North America and the spread of 'remote control' practices* (pp. 31–50). New York, NY: Taylor & Francis.

OPEN ACCESS

Civil society activists and vulnerability in South India: the relational politics of life history methods and development research

Matt Baillie Smith and Katy Jenkins

ABSTRACT

Life history methods are gaining popularity in Development research, linked to attempts to capture narratives marginalised by dominant accounts of Development. In this paper, we reflect on using life history methods with NGO activists in India. We explore how this approach led us to develop particular understandings of the participants as 'vulnerable', and the implications of this for the research process and the knowledges it produced. We explore how activists' individual biographies were interwoven with institutional narratives, complicating but also enriching our understanding of activists' experiences of Development. Secondly, we analyse the relationality of our subjects' vulnerability and our own positionality as global North Development scholars. We reflect on how our engagement with Development actors we consider as vulnerable takes place through and against the relational histories and presents that brought us together. We explore the implications of this for the ways the research created both discursive and physical spaces for meeting and talking, and what this means for our approach to vulnerability. This requires an uncomfortable acknowledgement that Development research may reproduce vulnerabilities, even as it seeks to challenge them. The paper contributes to broader theorising of vulnerability, recognising vulnerability as embedded in the relationalities of the research moment.

Activistes de la société civile et vulnérabilité en Inde du Sud: politiques relationnelles des méthodes d'histoire de vie et de la recherche du développement

RÉSUMÉ

Les méthodes d'histoire de vie sont en train de gagner en popularité dans la recherche du Développement, liée aux tentatives de saisir des discours marginalisés avec les comptes rendus dominants du Développement. Dans cet article, nous examinons l'utilité des méthodes d'histoire de vie avec les activistes d'ONG en Inde. Nous explorons comment cette approche nous a menés à développer des vues particulières de participants comme « vulnérables », les implications de cela pour le processus de recherche et les connaissances qui en ont découlé. Nous explorons comment les biographies individuelles des activistes étaient entrelacées avec les discours institutionnels, compliquant mais aussi enrichissant notre compréhension des expériences de Développement des activistes. Deuxièmement, nous analysons la relationnalité de la vulnérabilité de nos sujets et de notre propre positionnement en tant que chercheurs mondiaux du Développement du Nord. Nous réfléchissons sur la façon dont notre engagement avec les acteurs du Développement, que nous considérons comme vulnérables, se passe à travers et contre les histoires relationnelles et les présents qui nous ont réunis. Nous explorons les implications de cela sur les manières dont la recherche a créé des espaces à la fois discursifs et physiques pour se rencontrer et parler et ce que cela signifie pour notre approche de la vulnérabilité. Cela implique de reconnaître avec une certaine gêne que la recherche du Développement risque de reproduire les vulnérabilités, alors même qu'elle tente de les défier. Cet article contribue à une théorisation plus large de la vulnérabilité, en reconnaissant la vulnérabilité comme incorporée dans les relationnalités du moment de la recherche.

Los activistas de la sociedad civil y la vulnerabilidad en el sur de India: la política relacional de los métodos de historia de vida y de la investigación para el desarrollo

RESUMEN

Los métodos de historia de vida se están haciendo cada vez más populares en la investigación para el desarrollo, vinculados a intentos de capturar narrativas marginadas por informes de desarrollo dominantes. En este trabajo, se reflexiona sobre el uso de métodos de historia de vida con activistas de ONGs en la India. Se explora cómo este enfoque ha llevado a desarrollar una comprensión particular de los participantes como 'vulnerables', y las implicaciones de esto para el proceso de investigación y los conocimientos que produjo. Se explora cómo las biografías individuales de los activistas se entrelazan con las narrativas institucionales, lo que complica y también enriquece el entendimiento de las experiencias de desarrollo de los activistas. En segundo lugar, se analiza la relacionalidad de la vulnerabilidad de nuestros sujetos y nuestra propia posicionalidad como académicos globales de desarrollo del norte. Se reflexiona sobre cómo se lleva a cabo el compromiso con los agentes de desarrollo que se consideran vulnerables a través y en contra de las historias relacionales y los presentes que nos han unido. Se exploran las implicaciones de esto para las formas en que la investigación creó espacios tanto discursivos como físicos para reunirse y hablar, y lo que esto significa para nuestro enfoque de la vulnerabilidad. Esto requiere que se reconozca de manera incómoda que la investigación para el desarrollo puede reproducir vulnerabilidades, incluso cuando busca desafiarlas. El documento contribuye a una teorización más amplia de la vulnerabilidad que reconoce a la vulnerabilidad como incrustada en las relacionalidades del momento de investigación.

Introduction: vulnerability and development

Vulnerability is a central organising concept in the complex web of transnational practices, knowledges and representations that make up the global Development industry. The Development sector has been largely predicated on the idea of the existence of the vulnerable 'other' – the poor, marginalised, displaced resident of the global south to whom Development is 'delivered' in order to address their vulnerability and perceived 'lack'.[1] Bankoff

(2001) articulates this in terms of the way in which Development discourses, particularly around natural disasters, serve to render the non-Western world precarious and 'unsafe'. Ideas and scales of vulnerability are key features of Development strategies, evidenced in, for example, the targeting of extreme poverty, a focus on communities most exposed to climate change or projects supporting groups whose gender, identity or ethnicity produce particular exposure to structural violence. In this context, significant bodies of scholarship and policy analysis have addressed the multiple vulnerabilities produced through global inequality and targeted by the development industry (e.g. Naudé, Santos-Paulino, & McGillivray, 2009; UNDP, 2014), even if the word vulnerability has not always been the dominant frame. A key concern within these debates and wider critiques of capitalist development and Development interventions has been a vulnerability defined, in Liamputtong's (2007) terms, by people's relative invisibility and their 'marginality [and] lack of opportunity to voice their concerns' (p. 7) (e.g. see debates on participation (e.g. Cooke & Kothari, 2001) and feminist scholarship (e.g. Gluck & Patai, 1991)).

Representations of individuals and communities easily recognisable as vulnerable are also central in strategies to mobilise public and political support and raise funds for development. Through events such as Comic Relief and wider charitable appeals and campaigns, notions of vulnerability, particularly allied to the iconography of the child (Manzo, 2008) sit at the centre of popular cultural and emotional framings of global inequality and action to alleviate it. Such narrations of vulnerabilities in the global South do not necessarily challenge power or provide meaningful or productive accounts of marginality, and may even achieve the opposite, portraying poor people's lives as 'a permanent emergency' (Bankoff, 2001, p. 25).

A focus on the 'vulnerable' as a key subject of Development interventions is critically important, particularly in the context of the structural violence of neoliberalism and the growing gulf between rich and poor at local and global scales. But there is also a need to destabilise how vulnerability is constructed only as an 'object' for, and hence separate in some ways from, Development, since this risks reproducing popular and policy mainstreamed accounts of Development as an uncontentious form of rescue for the poor. As a professionalised and neoliberalised Development industry works transnationally to mobilise and legitimate particular knowledges, actors and authorities (Baillie Smith & Jenkins, 2012), we argue that attention is needed both to the vulnerabilities this can produce, and the challenges of researching them. In this paper, instead of focusing on the ways the Development industry addresses (or not) the needs of vulnerable 'beneficiaries', we consider the vulnerabilities the industry can itself produce. Our research uses biographical methods to explore the less visible and clear-cut vulnerabilities faced by civil society activists in South India – vulnerabilities which relate to NGO activists' structural location within the Development industry, and, as part of this, their exclusion or temporary access to the global civic spaces in which development professionals and their knowledges are legitimated. The research aims to foreground the ways in which Development impacts on individuals over time, and in ways that are outside an accustomed focus on 'beneficiaries', seeking to move outside the focus on Development's programmed effects, and to understand how Development as a complex of activities and relationalities produces vulnerabilities amongst those whose actions, relationships, trajectories and knowledges make up the industry. Our use of the term 'Development' reflects Hart's (2001) definition of 'big D' Development as a set of intentional processes – 'a post-second world war project of intervention in the "third world" that emerged in the context of decolonization and the cold war' (p. 650).

We begin by exploring biographical methods and their relationship to Development scholarship, before introducing ourselves and our research. We then go on to critically analyse our understandings of the activists as vulnerable, reflecting on the interviewees' blurring of the personal and the institutional in their accounts. In discussing the challenges these narratives posed to our initial assumptions, we analyse the questions this raises about our construction of the research 'subject', and how their vulnerability can be understood. We go on to link this to ideas of relationality and a consideration of the discursive and physical spaces constructed during the research, reflecting on the degree to which we risked reproducing as much as contesting the relations that produced the activists' vulnerability.

Biographical methods: unsettling dominant narratives of development?

Life history or oral history approaches focus on eliciting the narratives of individuals as they reflect on their life, usually in the form of an extended interview(s) in which the interviewee is given the space to speak freely, foregrounding events and experiences that are most salient to them. The biographical accounts produced are not 'factual' representations but capture the particular subjectivities of the interviewee at a particular time, providing an inevitably partial account that reflects the narrator's present circumstances as much as their past (Harding, 2006; Miles & Crush, 1993). As part of this, they are also relational in that they are produced through an interaction between the researcher and the researched.

Oral histories have been extensively used by historians and anthropologists as a method of recording ordinary people's accounts of the past, particularly in reclaiming community histories of struggle in the North. However, their emphasis has been on documenting individual accounts 'as supplementary sources of data or fact' (Miles & Crush, 1993, p. 92) In contrast, feminist researchers have adopted life history techniques not simply as a means of recording the lives of ordinary people but as part of a broader political project to give voice to marginalised women in the global North and South (Gluck & Patai, 1991). This notion of biographical research as emancipatory for vulnerable participants has been taken up across the social sciences in recent years (Harding, 2006). Harding (2006), citing Rustin (2000), suggests that this growing emphasis on individuals needs to be located in relation to the 'late modern focus on the individual and reflexivity' and the emphasis on the 'subjective and the cultural'. This focus on the self is also part of a move in Development scholarship to pay 'closer attention to development's routines, practices, and subjectivities' (Lewis & Mosse, 2006, p. 6), underlining the extent to which uncovering private stories and personal networks is central to developing more nuanced understandings of macro-level processes of social and economic change (Larner & Laurie, 2010).

Our research on activists in Peru and South India has explored how activists' professional and personal biographies reveal the operation of power in global civil society and the aid industry, and the ways this may be both constraining and enabling (Baillie Smith & Jenkins, 2011, 2012). We have shown elsewhere (Baillie Smith & Jenkins, 2012) how the Development knowledges and practices of the activists we discuss in this paper are produced through their negotiation of personal and career aspirations, commitments to their 'home community', strategic openness to donor organisations, exclusion from global civic spaces and changing regional policy. Through this, we can begin to see how subjectivities are a critical part of making sense of how Development is done in particular moments and places, and can understand how Development can work to marginalise and produce vulnerabilities over time.

Our use of biographical methods started with a commitment to promoting more just forms of development by foregrounding marginalised voices and knowledges. But such aims are not straightforward, resonating with much critiqued 'participation' in Development (Cooke & Kothari, 2001), and other attempts to 'give voice' in situations of continuing unequal power relationships where particular subjectivities may be being produced as much as liberated or empowered. For example, Lind (2003) argues that although the publication of Domitila de Chungara's life history as an activist in a Bolivian tin mining community gave her international prominence for several years, her life has changed little in material terms and she 'continues to live in deep economic poverty' (p. 233). Nevertheless, such activist biographies have been widely used in documenting everyday narratives of struggle in the lives of activists (particularly women) and marginalised communities in the global South, especially in Latin America (see, amongst many others, Burgos, 1994; Viezzer, 1977), and are perceived to provide scope for the reader to make sense of macro-level processes of social change through the lens of individual experiences. However, as Miles and Crush (1993) and Harding (2006) recognise in relation to biographies more broadly, such activist testimonials inevitably provide a partial and subjective account of the events narrated, and are not objective historical truths, but dependent on the cultural, social and historical context in which they are constructed, their perceived purpose and intended audience (Smith, 1999). Shank and Nagar (2013) reflect on the complexities of this in relation to Nagar's work with the Sangtin writers, shedding light on the ways in which processes of autobiographical storytelling involve 'a constant negotiation of vulnerabilities' (p. 98), as particular elements of individuals' lives are brought to the fore, or are kept hidden, depending on the socio-political context, the audience and the purpose that the narrative is to be put to.

Biographical approaches have not been widely used in critical development research, beyond a few notable exceptions (Lewis, 2008; McKinnon, 2007; Miles & Crush, 1993; Slim & Thompson, 1993; Yarrow, 2008). However, as McKinnon (2008) observes:

> … processes of subject formation are also processes in and through which dominant discourses may be reconfigured, shifted or combated. By examining the daily struggles of development professionals, it becomes possible to see clearly the productive possibilities that may emerge. (p. 290)

Our research began from the premise that life history research could provide an opportunity for less prominent Development actors to foreground their knowledge and expertise as they make sense of their past and present experiences. We hoped that it might also provide a space for them to reflect on how they position themselves both within and against dominant narratives of Development, challenging the assumption that development expertise is only provided by Northern professionalised Development workers.

Situating ourselves and our research

In 2009, we conducted 15 life history interviews with NGO activists in and around Madurai and Pondicherry, South India. The research aimed to capture the diverse trajectories of Development activists in local South Indian NGOs and the multiple ways they engage with, or sometimes are excluded from, the global Development industry, and how these processes and practices shape their cosmopolitan dispositions over time and space (Baillie Smith & Jenkins, 2012). We were particularly concerned with understanding the subjectivity of actors largely neglected within most accounts of Development and often marginalised from decision-making processes, despite rhetorics to the contrary (Baillie Smith & Jenkins, 2012) – global South NGO activists whom we characterise as 'intermediate', in that they fall between the grassroots, on the one hand, and the professionalised and globally networked on the other. This intermediate status is, as we will go on to argue, of central importance in shaping their vulnerabilities within the Development sector, as well as how they negotiate these vulnerabilities.

Recent years have seen a growing interrogation of the relationships between cosmopolitanism and Development (see, amongst many others, Baillie Smith, Laurie, Hopkins, & Olson, 2013; Berry & Gabay, 2009; Held, 2006; Kothari, 2008). Most relevant here are analyses of subaltern cosmopolitanisms (e.g. Datta, 2009; Gidwani, 2006; Kothari, 2008), emphasising how cosmopolitanism can be strategic, practical and political and rooted in the everyday. In this light, our work interrogates how a strategic openness to difference influences the negotiation of global civic spaces by actors who are neither 'subalterns' nor the elite mobile citizens of International NGOs (INGOs) (Baillie Smith & Jenkins, 2012). We argue that being cosmopolitan can be understood as forming part of the repertoires of people negotiating vulnerability, resonating with Wiles (2011) assertion that whilst '[v]ulnerability may be conceptualised as fragility and (or) weakness, [...] it could also be conceptualised as openness, susceptibility, and receptiveness' (p. 579).

Working with a local Development NGO with whom one of the authors had a long-standing connection as both an academic and, previously, a development practitioner, access was obtained to activists from a range of NGOs, all of whom had been involved in Development and what might broadly be described as 'activism' in various guises for at least 20 years. We are therefore specifically focusing on older activists, individuals in their 40s and 50s. Whilst we acknowledge that the notion of an 'activist' tends to be rather slippery (Maxey, 1999), it is somewhat less so in the Indian context where the term is more readily used to describe those individuals who actively and consciously engage in, and are embedded within, networks and processes aimed at fomenting social and political change, including but not limited to grassroots organising, social movements, NGOs, community volunteering and faith-based organising. Our research participants were therefore individuals identified by the partner NGO as prominent local 'activists', all of whom also self-identified as activists in the narratives they related. For many, their NGO work was only one element of their activism, which was characterised by a multiplicity of everyday practices (Jenkins, 2014; Maxey, 1999). We included activists working in organisations with a range of development interests, from women's activism and child labour, to disaster relief and Dalit activism, aiming to capture the different ways cosmopolitan subjectivities might be shaped through particular areas of expertise and knowledge and how these were constrained or opened up by the Development industry.

Interviewees were not selected as representatives of a particular NGO but for their individual trajectories – we aimed to capture their experiences of activism through, beyond and across particular organisations. For example, the activist biography of Simon recounts his early participation in anti-China street protests at the age of 8 or 9, his heavy involvement in student politics at college, how he later combined his activism with a job as an academic, and most recently his role in helping run a small NGO. Nevertheless, despite our initial focus on individual trajectories, the blurring of individual and institutional histories subsequently became a key factor in our analysis, as we explore below. Interviewees were not selected due to a perception that they were 'vulnerable' but rather the notion of vulnerability emerged later as a way of making sense of their precarious position within the Development industry.

Interviews were conducted by both authors at various locations; in some cases we visited activists' organisations, at other times they came to where we were staying, something that for one activist was explicitly related to concerns about being made vulnerable through being publically associated with international visitors (Baillie Smith & Jenkins, 2012). We were alert to this concern throughout the fieldwork (and beyond), and this made us particularly sensitive to issues around securing anonymity of both organisations and individuals.[2] Like McKinnon (2007) in her work with Development professionals in Thailand, we were specifically concerned with individuals' activist biographies rather than entire life histories; all interviews started with the question 'Can you tell us about how you became an activist and how your experiences have changed during your time as an activist?' Interviews were tape recorded and mostly conducted in English, with transcription (and some translation from Tamil)[3] undertaken by our partner NGO.

Understanding intermediate NGO activists' vulnerabilities

Our focus was on the position of actors whose intermediate status within Development produced particular subjectivities. We conceptualise this intermediate status in terms of their positioning in-between key foci of Development policy, research and practice, simultaneously negotiating local, national and transnational discourses and spaces of development and acting as 'brokers and translators' (Lewis & Mosse, 2006). This intermediate position produces vulnerabilities and exclusions which have not been readily identified in the literature (Baillie Smith & Jenkins, 2011, 2012); their position at the intersection of multiple development arenas means that they are much less researched than those situated firmly at the grassroots or within the global sphere. Their intermediate status produces a less immediately apparent vulnerability that does not easily 'fit' established work on Development or on vulnerability. Further, these activists' positions within NGOS often automatically frame them, in the Indian context, as privileged, resonating with Chatterjee's (2008) identification of Indian civil society as the preserve of the middle class.

The vulnerability of this group particularly stems from the insecurity they face as workers in the rapidly changing context of the Development industry in India, where priorities and foci shift frequently, most recently in relation to the reduction of aid to India:

> In the past ten years the government would mark an area as a developed area but actually lack continues. The government data would point out that education has been achieved, infrastructure is available. The government announced like that. International donors would think that Tamil Nadu is a fully developed state. They would move to another area. In this situation, funding is reduced for Tamil Nadu and Pondicherry. No funding is given for this area. (Rajesh)

This insecurity was articulated by several participants, who particularly highlighted the rapid influx of resources for disaster relief immediately following the 2004 tsunami, but also its equally rapid disappearance and the consequences of this for small-medium NGOs and their workforce. The anxieties this generated are highlighted by two activists from Pondicherry, whose organisations grew massively following the tsunami:

> One of the things I always find a threat, irrespective of donor or which NGOs is they all have a very time bound approach. (…) And all of sudden at the end of three years you are told that thanks for your hard work and all. And that man they don't get a job. Here, for the last three months, I've been seeing, almost every 30th of month, some long faces. When I ask them, they say that

> we don't know what will happen today or in the evening. And what happens is a minimum of 10 or 15 people are losing their job every month. No project and no funding is coming. So what is going to happen? (John)
>
> The flight called 'Pondicherry Action' is flying in a very high level and it is slowly coming [down]. Still it has not landed. It would take another three or four months. After tsunami, we had around 250 staff. Now we have 40. Slowly we are coming down, completing the projects, handing them over to the community, reporting to the donors ... (Paul)

Understanding the position of these intermediate activists is important because they embody wider structural vulnerabilities inherent in an increasingly neoliberalised aid model. At one level, their vulnerability is a function of professional engagement in Development work with vulnerable and marginalised groups, at a scale that is of diminishing interest to donors and the aid industry. But their vulnerability actually comes less with a particular status, and is as much to do with their intermediate and in-between position – their movement between a professional and personal vulnerability, and the ways these intersect, overlap and flow into and out of each other. The situation of John, who had previously worked for a large INGO and who had now returned to work for a smaller organisation in Pondicherry illustrates the challenge of defining vulnerability. He spoke in strong terms about his lack of pension and support for old age as a result of poor conditions linked to working for an INGO, conveying a strong sense of vulnerability and concern. He is at once a 'Big Man' (Yarrow, 2008, p. 338) of the NGO world, and apparently relatively comfortably off, but he is simultaneously vulnerable as the aid industry moves on and employment opportunities disappear. His commitment to the INGO is now having impacts on his family as he reaches retirement age:

> And now I am 56. What is next? Now the next three or four years' time I have to retire and what's going to be after my retirement? Of course there are people who continue to work as consultants. But I am tired. It's nearly 35 years now. I have been travelling travelling and doing all sorts of jobs. I don't think that my health is that good enough to manage me going as a consultant. (...) Retirement means really a retirement. The time I have denied for my family all these years. (John)

This may not then be the 'normal' vulnerability of aid and Development, and John's status and apparent wealth place him in a far less vulnerable position – as far as we could tell – to the majority of the Indian population. But if we were to frame this in the context of current labour relations debates in the UK and elsewhere, around issues such as casualisation, temporary and zero hours contracts, and the decline of workplace pension schemes, we would understand him as vulnerable.

The difficulty of finding the individual and the blurring of the personal with the professional and institutional, reveals the activists as simultaneously embodying multiple vulnerabilities and privileges. This situation provides a useful lens on the interplay of the professional and the personal in understanding vulnerability, but also presented a 'challenge' given our interest in exploring a less institutionalised account of civil society and Development: a seeming reluctance on the part of the interviewee, or failure on our parts as interviewers, to elucidate activists' individual narratives. Despite different prompts and sub-questions, when asked about their particular story of activism, the majority of interviewees narrated the stories of the organisations they were working for or had worked for. Personal and organisational histories were blurred and difficult to disentangle (Baillie Smith & Jenkins, 2012); the 'I' we had thought we were interested in was consequently often consigned to a background role, with attempts to probe further often recast in terms of an organisational identity and history. On some occasions, interviewees mapped their history and that of their organisation onto well-established genealogies of development practice, referencing key approaches such as 'basic needs', resonating with Yarrow's (2008) highlighting of the performative aspects of narration. However, in doing so, the interviewees also provided an established – if contestable – narrative of development which then displaced the details of their subjectivity within that narrative (see also McKinnon, 2008). This can be seen in the response of one interviewee to the question of how the experience of being an activist had changed over the past 20 years, which he recounts as a series of shifting development priorities and initiatives:

> It took me those [first] three years to build some kind of a platform to stand upon with [the NGO]'s ideology. (…)I slowly started the community mobilisation and promoting awareness. (…)Then we identified the real problem of the community in our operational area. Then we formulated the projects for these people. (…) Within two years we stopped the activities. (…) and then the government asked us to take up the women development programmes for the first time. It was a good opportunity as I got recognition from the government. So I got this opportunity for collaboration with them. Our task was to converge all the resources of the women groups like Micro-credit and Self-Help Groups towards the under-privileged community of women. (…) In the beginning we started social empowerment. Now we have shifted to economic development. (Paul)

We initially felt that what was missing was the 'particular' of the stories of the individuals with whom we were talking. This can be seen in interviewees' discussions of their overseas experiences. Participation in international meetings or visits to Western Europe or the USA was defined by most interviewees as key moments in their careers and personal development. Whilst interviewees did reflect on what opportunities these international engagements opened up for them and their organisations, when we asked about personal experiences of such opportunities, we were unable to elicit further detail. This is illustrated below in the separate responses of two activists to our question about whether they had faced challenges interacting in the transnational spaces of which they spoke (coincidentally both in Brazil), the World Social Forum (WSF) in the case of Matthew and an NGO training programme in the case of Jagdish:

> There are different types of learning. Mostly positive learning. The situation is totally different. Land holding is totally different there. In India, we are talking about small and marginal farmers. Here in India, marginal farmers means below 2.5 acres. Up to 5 acres, small farmers. But in Brazil, even a single farmer has around two thousand acres or five thousand hectares of land. Totally different situation. So we could not take whatever information we could collect there to India. We learnt and just modified them to fit the local conditions. (Jagdish)

> With the World Social Forum, the important thing is to bring all the activists from different corners of the world. That itself is something meaningful. But how far, we can articulate to become a force, becomes difficult as such. But this time Belem, Belem is very far from here. Brazil at another end, it was not a place for global meeting [the World Social Forum]; it was in the Amazon forest. So, from India there were only a few participants because the travel itself is very costly. But the important thing I saw is, Amazon forest is kind of the lungs of the world. Lungs are the oxygen producing place. So that forest has to be protected. The meaning for conducting this programme there is for that purpose. The forest there is protected and also the indigenous people. (Matthew)

Rather than reflecting on any personal challenges, both responses are instead more generic and focused on the knowledge they gained for their organisations from the event. Jagdish's discussion is framed by the challenges for his NGO in applying the knowledge offered by such transnational experiences, whilst Matthew reflects on the choice of location for the WSF and how that framed the ideas he took away from the event back to his organisation. Questions asking 'what it was like' or 'how it felt' to be at such an event in terms of day-to-day issues tended to be met with confusion; personal experiences were thus obscured or downplayed. We felt as if individual biographies remained concealed, prompting us to reflect on whether we were asking impertinent questions, including ones which were perceived to question the authority, expertise or skill of the interviewee, rather than opening up spaces for their expertise to be recognised.

Reflecting on the challenges of disentangling the personal and institutional, highlights issues of trust, openness and vulnerability and the need to interrogate the implicit assumption of researchers that interviewees are willing and able to reflect openly on themselves and their experiences (Lewis, 2008). As Yarrow (2008) observes, 'it is a mistake to assume that people everywhere regard biographical information as interesting or revealing in the way that many in the West imagine' (p. 336). Identifying the activists' responses as problematic because we felt their individual narrative was missing is to effectively construct a subjectivity into which we expected the activists to fit, itself potentially invoking or producing a further sense of vulnerability or exposure:

> ... biographical research (in both social science and oral history) tends assume a biographical 'I' who is typically self reflexive and prepared to speak publicly about his/her experiences. This biographical 'I' is one who is expected to demonstrate a specific form of rationality: drawing lessons from the past and using these to inform future plans. This subject, especially in oral history, is positioned as potentially empowered through the narration of his/her past experience as it becomes visible to and acknowledged by self and others. The questions that follow are: how do interviewees take up, rework or refuse the positions created for them in advance by researchers? To what extent does the interviewee become the 'I' the researcher expects and wants him/her to be? And, if biographical subjects are assumed to have agency, do they demonstrate the sort of agency and rationality expected by the researcher? Or, do they do it differently? (Harding, 2006)

The intermediate status of the activists we interviewed, and their organisations, is again important to consider here. Their position working at the interface of local, national and global processes and institutions may accentuate their actual or perceived vulnerability and thus their desire and ability to reflect openly on their position within macro-level processes. This is particularly pertinent given that researchers conducting biographical research with civil society actors who occupy less precarious positions have not encountered such reticence to recount personal narratives (Lewis, 2008; Mawdsley, Townsend, Porter, & Oakley, 2002). Additionally, it is important to situate our interviewees' responses in the broader context of diminished international donor engagement in South India, particularly shaped by the government prioritisation of poverty reduction in other states, and the challenges this presents for small NGOs and their continued survival. Consequently, and linked to the ways we were introduced to each other (discussed below), activists' approaches may also have been strongly shaped by the particular challenges faced by their institutions, and a familiarity with presenting narratives that might serve to generate much needed income. Here, again we see the blurring of individual and institutional narratives, and indeed vulnerabilities, with the fortunes and livelihoods of individual activists closely intertwined with those of their organisations (Baillie Smith & Jenkins, 2012):

> For me, starting the organisation was very difficult. Moreover finance is an important component. (...) I struggled for three years. I travelled to many places on a bicycle and I didn't even take food for one or two days together. (Paul)

> So this is the time that I started feeling that my friends ... they all now are in very good positions. Some of them are now even become chief engineers and some of them are professors in Ana University. And whenever I look at them or I meet them I feel regret that I have chosen the field of NGO sector. And ultimately landing up in the problem of insecure feeling. (John)

Unpicking this blurring of personal, professional and organisational narratives requires us to re-frame our initial concern at the apparent absence of personal reflections, and instead recognise the particular ways in which we were constructing the participants' vulnerabilities in the first place. Life histories are relational – they develop between researched and researcher – pointing to the need to understand how the research itself and we, as researchers, fit into the complex ways in which vulnerabilities are shaped through engagement in the Development industry. Grappling with these challenges led us to explore the ways multiple relationalities shaped the discursive and physical research spaces of the research, working through and against Development and creating instability and contestation around the doing of the research.

Vulnerability, relationalities and life history research

If we are to argue that activists' cosmopolitan subjectivities are negotiated through and against Development and the transnational encounters it produces (Baillie Smith & Jenkins, 2012), then we need to acknowledge and interrogate how such encounters also take place in our research, and how we are also involved in that process of negotiation. Pillow's notion of pursuing an 'uncomfortable reflexivity', 'a reflexivity that seeks to know while at the same time situat[ing] this knowing as tenuous' (Pillow 2003 in Nairn, Munro, & Smith, 2005, p. 222) is useful here in conceptualising the interview as a productive relational space. In this section, we consider the multiple and overlapping relationalities through which our engagement with the activists took place. We particularly reflect on ambiguities and contestation in the construction of the discursive and physical spaces of the research, and reflect on how our research became part of the relations and structures that produced the activists' vulnerabilities even as it sought to challenge them. In this way, we explore how, in Harding's (2006) terms, the 'I' of the activists was constructed.

Conceptualising the interview process as relational acknowledges power relations, but also demands we bring our 'Northern' selves into the process, contra the 'othering' histories and traditions of much Development scholarship. As well as drawing on the extensive feminist theorising of positionalities within the interview space itself (Alcoff, 1991–1992; England, 1994), this involves recognising that the 'stories' that we are told will never be 'identical to that which the same narrator would give to another person' (Devereux cited in Behar, 1996). This underlines a need to write the 'self' in (Cotterill & Letherby, 1993) as part of recognising that critical Development researchers – like the Development industry – work 'in' and 'against' (Kothari, 2005, p. 62) Development's histories and dominant narratives. It also reminds us that we cannot escape the relations we may want to problematise, but should reflect critically on how and where our approach actually reinforces such relations, where it may open up spaces to unsettle them, and when it may do both.

The critical capacity of our biographical methodology may thus arise as much in exploring and understanding the layers of intersubjectivity in the interview process as in the 'formal' data it generated. As Alcoff (1991–1992) reminds us: 'Who is speaking, who is spoken of, and who listens is a result, as well as an act, of political struggle. Simply put, the discursive context is a political arena' (p. 15). An emphasis on understanding life history approaches in terms of intersubjectivity would seem to fit our desire to pay attention to voice, positionality, and understanding subjects' vulnerabilities. However, we should also recognise the ways in which the positionalities of the 'researcher' as well as the 'subject' and the 'individual' are themselves multiple and fluid and co-constituted and negotiated in the research moment (Chacko, 2004).

Intersubjectivity focuses attention on the relationality of subjectivity. Whilst there are multiple definitions of intersubjectivity, rooted in diverse disciplines and traditions, we follow Gillespie and Cornish's (2009) inclusive definition of intersubjectivity as:

> the variety of relations between perspectives. Those perspectives can belong to individuals, groups, or traditions and discourses, and they can manifest as both implicit (or taken for granted) and explicit (or reflected upon) (p. 19)

At both the level of North–South encounters and the lived experience of being an NGO activist in South India working between different Development spaces, we can see the significance of intersubjectivity to development thinking and practice. Intersubjectivity is also critical to our conception of cosmopolitanism in terms of an openness to difference as this is practiced in the encounters fostered by Development and global civic spaces. How we were introduced to each other then becomes increasingly significant for defining the space in which our subjectivities interact, as individuals' histories, presents, engagements with, and resistances to, Development surface in the research moment. As Cotterill and Letherby (1993) recognise, 'All research contains elements of autobiography and biography, both intellectual *and* personal. Autobiographies and biographies not only record the life of one individual, they are in a very real sense documents of many lives' (p. 68).

Reflecting on the relationalities of the interview space, we now realise that it was significant that one of the authors' own and family history of engaging with India and with South Indian NGOs, particularly as a former 'practitioner', was used by our partner organisation as both an introduction and legitimation when we first met interviewees.[4] This then necessitated stepping outside the comfort and formality of a professional identity, and talking openly about the potentially contentious topics of his father's volunteering history in the 1960s and grandfather's colonial era presence, their legacies, and how these came together with a changing professional engagement with India and Development. This at one level problematised any presumed encounter between academics and practitioners that our interviewees may have anticipated. On the other hand, that the other (younger, female) researcher (who did not have this practitioner experience) did not receive this introduction but rather was introduced as an academic who researched activism, may also have complicated and confused the setting and understandings of our subjectivities; were we coming at this as academics or in solidarity as activists? Alternatively, did that activist history locate us within the elite and Northern-dominated global civil society we were seeking to interrogate? We could argue that the NGO history of one of us created an encounter framed by the very issues and discourses we were seeking the activists' perspectives on. So in establishing some 'common ground' and seeking to demonstrate our solidarity with the activists through the previous experience of one of us, we were also creating a research imaginary in which we were immediately located, placing the interviewee in a difficult position in terms of constructing a critical account in which they might feel we were ourselves implicated – as one interviewee commented, 'I do not know how to say this to you as an international person' (Joseph). As we sought to engage with the activists' subjectivities over time and across the spaces of Development, the construction of one of us as a *former* practitioner complicated the temporal situating of the research. As Chacko (2004) recognises, 'the manner in which [the researcher] engages differences and commonalities and represents them through field research practices has contextual and temporal underpinnings' (p. 54). We effectively blurred and destabilised the 'when' and the 'who' of one of our positionalities just as we sought to investigate the activists through these concepts. In doing so, we were ourselves reflecting the importance but also complexity of changing relations to Development over time, but also confirming that these are not discrete moments, but flow in and out of each other. This then illustrates the complexity of defining the activist 'I' of the research as it emerged through relationships within and beyond the research moment and over time.

Issues of language and meaning are particularly key in situating ourselves and our research participants, and in structuring the discursive space of the research encounter. Most interviewees chose to speak English during the interview, whilst a few spoke Tamil with the interview conducted using a translator. Whilst qualitative research always involves a degree of interpretation of meaning, even when using a single common language (Smith, 1996), issues of understanding are particularly pertinent when working across and between several languages. Liamputtong (2010) notes that 'concepts can move problematically across cultures' (p. 146), and this is particularly relevant when participants are speaking a language that is not their first language, as was the case with this research.[5] One of our participants, Kumar, who chose to speak Tamil during the interview, explicitly reflected on this dynamic in his interview, underlining that his aim was to share his thoughts as clearly as possible, and that he could do this better in Tamil:

> I want to add something. English is one language. I feel problematic to convey in English to you. So I need a translator. This is only, I mean, professional. I don't need to take risk by conversing in English. My aim is to convey the message to you. (Kumar)

Such issues around language and meaning are important to consider in thinking through the power relations embedded within languages, particularly in the context of aiming to create open research encounters, where participants are able to voice concerns and vulnerabilities. Of relevance here is the status of English as the lingua franca of the Development industry – the language of professional interactions in which NGO activists need to be competent in order to navigate the development sphere and interact with donors and Northern Development professionals (Batliwala, 2002). This is not plain English, but rather a language of buzzwords and 'donor speak'(Mawdsley et al., 2002).The extent to which interviewees situated us as part of these discourses is important; on some occasions, despite having explained the nature of our research, participants did ask if we could help provide project funding, firmly locating us within the very processes we wished to critique. Here, we can also perhaps see how a past location within the Development industry as a global North professional, runs into and creates further layers of meaning in the current research moment and our positioning in relation to Development. Perhaps then, we should not be surprised that conducting interviews in English led interviewees to use such 'donor speak' to provide the linguistic resources for articulating their experiences and structuring their narratives. This also contributes to the obscuring of individual trajectories, experiences and subjectivities, through generating a professionalised rendition of their activism, wrought in the language of 'capacity building', 'evaluation' and 'projects', and produced through, rather than contesting, dominant Development discourses. On reflection, we recognise the same vulnerabilities that activists experience in their dealings with Northern NGOs and development workers, are simultaneously reproduced in the interview context by the use of English language. This might be understood as reinscribing historical and contemporary Development relations (Temple's (2002) 'linguistic imperialism', cited in Liamputtong, 2010), unintentionally creating a space circumscribed by the hegemonic Development relations we sought to critique.

This discursive space of the research encounter was also shaped by the physical spaces in which the interviews were conducted, usually the offices and fieldsites of our participants. Despite aiming to put participants at ease by interviewing them in a familiar setting, these spaces again perhaps limited the activists' abilities to openly express critique of the Development industry, or to reflect on how they negotiated their precarious position within it. Conversely, one participant specifically requested that he come to our hotel to be interviewed, creating a space in which he felt able to articulate the vulnerabilities he experienced as an activist involved in raising awareness about child labour in the matchstick industry:

> It is definitely good to have a relationship that is built [on international] links that is the positive side, but the negative side is that the individual who works [in this way] comes under the scanner of the international bureau, enforcement agencies and they may try to give kind of a pressure, watching them every moment, so the pressures are high and that's the negative side. (Vinesh)

This participant's request to meet at our hotel was in order that he not be observed meeting us, which he felt would expose him to greater scrutiny, underlining the potential that as Northern researchers our actions may reinforce vulnerabilities and dominant power relations. As Geiger (1986) notes, 'informants run risks that may not be immediately recognizable or even preventable' (p. 350). The powers and vulnerabilities of researched and researchers do not sit outside of these spaces, but rather are integral to them. The activists' subjectivities and vulnerabilities thus themselves shape, and were evidenced in, their engagement in, and openness to, our research. At the same time, as we worked to reconfigure our ideas of the individual we were pushed to reflect on our own subjectivities, both in the research moment and in the subsequent analysis. In particular, our commitment to understanding and highlighting hidden and alternative voices in development, and our own positioning within a mobile global elite and use of this mobility for our research ends, were all challenged.

As well as considering the construction and negotiation of intersubjectivity in the interview context, locating researchers and researched in their social, cultural and historical contexts may provide additional ways of thinking through the relationalities constructed in the research moment and beyond. Perhaps most significant is the very notion of subjectivity and the ways this concept is interpreted and defined. In this research, we can partly situate the challenge of identifying the personal by recognising interviewees' responses as grounded in a less individualised and commodified society than our own. As Arnold and Blackburn (2004) observe, 'Life histories in India do not necessarily conform to Western conventions and modes of expression (some do, many don't), nor should one expect to find the peculiar forms of individualism that emerged in the West replicated in India' (p. 3). So the intersubjectivity of our interviews is then produced through the interactions of our (ambivalent) rooting in an increasingly neoliberalised context which celebrates and promotes individual autonomy and rationality, and the location of the interviewees in a context where ties of community and ideas of collective action are seen as more significant. As two interviewees reflect:

> We don't have that [individualism] … I just give a simple example. Ask any Indian whether he maintains his journal or diary. They may not. It is very difficult, even for me. I just tried as an experiment. Because we don't exist as individuals. So when can you write a diary? When you are conscious of yourself. When you are sure of your privacy. So the concept of privacy is not there … You can't be an individual in that sense, which is all about your experience and expression of your experience. (Simon)

> Individual is part of a bigger system. (…) In the Western context the individual is important. Whereas here it's the family, school, college, organization … (James)

Interviewees' cultural, religious or social identifications – such as being a 'dalit', 'Marxist', 'Gandhian' or 'brahmin' – also explicitly framed their narratives. Such identities shaped the ways that they and we perceived their vulnerabilities, and demonstrated a sense of identity and individuality that emphasised a 'constant interaction and negotiation' between collectivity and individuality (Arnold & Blackburn, 2004, p. 3). Activists identified particular notions of 'service' – which some of them linked to Brahminical traditions – as shaping their identification of their roles, histories and subjectivities, with these categorisations effectively displacing the focus on the individual and on individual achievement. A similar dynamic is also reflected in the work of Kagitcibasi (2005), who explores the intersections of autonomy and relatedness. In contrast to the Euro-American conceptual separation of independence and relatedness, Kagitcibasi (2005) posits that autonomy and relatedness can coexist in an 'autonomous-relational self', emphasising the interplay between collective and individual identities in the construction of a 'relational self', that particularly resonates with the Indian context and the comments of our interviewees. Our data reveal a temporal dimension to this interplay. A number of our interviewees located part of their activist identities in lamented histories of collectivist traditions and left politics in India, although such ties were recognised by interviewees as increasingly being undermined, leading some to use a temporal and generational lens to define their more collective activism in opposition to neoliberal and individualistic ideologies of younger activists:

> One problem is younger generation in India, I think more and more they are money centred (…). To make the younger generation get into social activism itself is a difficult thing. That is one thing. Secondly, social activists, now, I don't know, I can't say much because in my days, when I started it was a different ethos, there was something in the country, we wanted to get involved, so we took a lot of risks but nowadays we don't get that kind of set up. (Matthew)

> We used to go and sleep in the hut with the people, we had no problem, we ate whatever they ate, we went to struggle area. I mean we have to walk 18 kms. I had hurt my whole feet, I came back and I had to undergo treatment for nearly two months. So that is the hardship we went through. But this generation, (…) I see that they are not willing to compromise on their comforts. That is why, maybe I am not able to understand them fully but, this is what the reading is. They want the solution very fast, impatient. (Chandra)

Only in paying attention to the production of individual identities and subjectivities, as well as how the interview setting brings these into dialogue with our own subjectivities, can we begin to understand the issues around the activists' narrations of their own individual histories. This highlights a need to focus on how a set of tensions were played out, both historically through the activists' life histories, and in terms of their current experiences in negotiating their professional and personal identities and organisational trajectories, as well as in terms of the spaces which we and they were shaping in our interview interactions. As Findlay (2005) recognises, vulnerability 'is constructed in relation to the boundaries that distance "self" from "others"' (p. 433). Despite historically shaped collectivist orientations and ambitions, the activists' organisations, experiences and vulnerabilities were increasingly shaped by processes we and they were seeking to critique: the neoliberalisation of civil society and associated emphases on outputs, service delivery and particular modes of accountability that flatten out particularity in the name of certain 'universals'.

An example of this is some participants' reluctance, and often refusal, to sign the consent forms which our institution expected us to use. We can see this as suggesting a resistance to commodification and audit, as well as underlining their own feelings of vulnerability (particularly around possibilities for identification) and a broader distrust of signing formal paperwork, common to marginalised peoples. The consent forms also served to emphasise our own contradictory position both within and against such audit-driven processes. Indeed, we could argue that our focus on the activists as individuals also unwittingly exemplified the processes of individualisation that they and we were seeking to resist. At the same time, interviewees' use of organisational histories as a proxy for more individual narratives bears testament to the activists' continued capacity to work in and against the pressures we were trying to understand. This may not then be the writing out of the individual and the particular that we initially assumed. Rather, we can see it as reflecting a sense of collectivity and solidarity, and for some, a sense that their lives are intimately entangled with the organisation and do not have a separate trajectory, particularly if they have founded it:

> I relate with my organisation because I belong to that organization, I relate to the incidents and development that has taken place. So instead of myself that has done that ... I don't know. Maybe because you are involved in the organisation, and you become part of the organisation. (James)

> In our calendar, we have lost five years without even knowing that. So we had that type of busy work and running around here and there. We mostly lost on the family side in our personal lives and things like that. In the past four-five years, our main thinking was how to implement this programme in a responsible way. Because, suddenly there is so much of fund coming in and we cannot simply lose time in implementing this programme. We were putting in all our energy in carrying out this programme in a responsible manner. So there is this personal side – like we do not know how the last five years went. (Joseph)

Despite our initial emphasis on producing a less institutionally focused account of civil society, for the activists themselves, their commitments to social justice were in large part manifested through the interweaving of their personal and political biographies and a range of institutional and collective narratives. This suggests a need to understand our research as a dialogue across individual, professional and organisational histories on both sides, in ways that sometimes reflect, and sometimes work against, the processes of elite knowledge production and practices of cosmopolitanism that we were seeking to engage with. It also suggests the need to recognise that the spaces created by biographical research in Development, and the potential intersubjectivity that results, are likely to be messy and uncomfortable. However, following Kaldor's emphasis on civil society as a conversation (Kaldor, 2003, p. 160), such intersubjectivity may present the possibility of effecting change beyond simply producing elite academic knowledge. Rather than reflecting on whether our research process was flawed or failed to transcend a world order we wished to critique, a more productive approach acknowledges the ways 'nano level' encounters, which seek to generate critical knowledge, are not divorced from the unequal histories and presents which we and the interviewees are part of, but contribute to a continual process of joint negotiation. Shank and Nagar (2013) frame this in terms of the added layer of responsibility for researchers using storytelling approaches, creating a need to 'continuously renegotiate the multiple and fragmented vulnerabilities of activists' (p. 100). As Russell and Kelly (2002) also recognise:

> The concept of negotiation within the interview suggests that the process does not merely recount past events; rather, it constructs new stories out of the flow of information and interpretation of both participants. In this context, research results emerge from the interactive dialog that occurs between researcher and respondent as they position and shape the information ultimately captured in the text. (no page)

We therefore recognise the potential of life history methods in terms of their contribution to the co-creation of a space – by ourselves, our research partner and the interviewees – in which ideas, experiences and memories are shared and co-constructed (Slim & Thompson,

1993), something which distinguishes our approach from the dominant paradigms we were seeking to resist. Re-imagining the interview space as relational, cosmopolitan and co-constructed resonates with Wiles' (2011) association of vulnerability with openness and receptiveness. This reminds us of the extent to which creating these spaces also necessitates that as researchers we make ourselves vulnerable, through revealing our own life histories. This does not make the process of co-creation even or stable, and we have seen how differently constituted and understood vulnerabilities – and power – came together in and through the research process. But despite the challenges we initially encountered in making a life history approach work for our research and our participants, the spaces we created did enable us to capture the multiple vulnerabilities experienced by intermediate NGO activists, and how these often cut across individual and institutional boundaries.

Conclusion

This paper emphasises the importance of thinking about researching vulnerabilities in terms of an intertwining of the personal relationalities of the research moment, and the wider relationalities of Development in which we were both consciously and sometimes unconsciously caught up. This process is then shaped by our own particular perspectives, histories and biographies and their connections to those of our interviewees. Our life history approach captured both how activists' perspectives related to wider public discourses around Development, and also the ways in which they related to us as an audience. This is not to suggest a separation of these two elements, as we were also caught up in those same public discourses, which in turn shaped how we were understood as an audience. Our focus on intersubjectivity is therefore central to enabling a critical analysis of the knowledges produced by a biographical approach to Development research as it foregrounds the multiple relationships at play in the research process and research setting, and underlines the extent to which approaches that focus exclusively on the individual risk failing to capture the multiple relationalities through which vulnerabilities are constructed and contested. Through a focus on intersubjectivity and relationalities we can thus better understand the ways in which Development research may reinforce and become part of the structures and relations that produce particular vulnerabilities even as it seeks to critique them.

Nevertheless, we argue that despite the challenges of working across tangled individual, professional and organisational narratives, the life history approach we adopted provided an opportunity to generate a critical analysis of the multiple vulnerabilities experienced by intermediate Development activists, as well as underlining the ways in which researching vulnerabilities requires us to better reflect upon our own subjectivities and the often unanticipated ways these interact with those of our participants.

The paper therefore contributes to broader conceptualisations of vulnerability, in particular in relation to Development. We argue for the need to go beyond straightforward understandings of vulnerability as related to poverty and exclusion, and for notions of vulnerability in the global South to encompass actors whose precarious position within the structures of Development renders them vulnerable in unexpected and less visible ways. Although beyond the scope of this research, this positioning of intermediate NGO activists as vulnerable should not be understood as specific to the global South context or the Development sector. In the context of continuing austerity and insecurity, such notions of precarity and vulnerability may also map on to the life histories and experiences of third sector workers and civil society activists in the global North. Thus, we hope that this paper opens up new avenues of enquiry and methodological possibilities for thinking about how a life history approach enables ideas of vulnerability to be explored in diverse contexts and across individual and institutional boundaries.

Notes

1. See Escobar (1995) and Crush (1995) for a detailed critique.
2. For example, we have provided few details on the specifics of organisations that activists were drawn from, and have omitted information that might serve to identify individuals. In the field, we acted discreetly in relation to conducting interviews, and participants were able to choose locations where they felt safe and at ease.
3. Issues around translation are discussed below.
4. There are obviously strong connections here with the feminist literature on positionality and insider–outsider dynamics in research.
5. We are also cognisant of the 'Indianization of English' (Kachru, 1986), meaning we cannot assume an equivalence in how we and our interviewees understand questions posed in English and responses to them. Liamputtong (2010) and Twyman, Morrison, and Sporton (1999) also discuss the complexities of cross-lingual research and the use of translators.

Acknowledgements

We would like to thank the activists who took time to share their experiences with us, CEDAR, Madurai, for their support in carrying out this research, and Northumbria University's Faculty of Arts, Design and Social Sciences for funding the fieldwork on which this paper is based. This research emerged from the ESRC seminar series 'Activism, Volunteering and Citizenship' (RES-451-26-0561), and we would also like to acknowledge the contribution made by the participants in shaping this research.

Disclosure statement

No potential conflict of interest was reported by the authors.

Funding

This work was supported by Northumbria University's Faculty of Arts, Design and Social Sciences.

References

Alcoff, L. (1991–1992). The problem of speaking for others. *Cultural Critique* (Winter), *20*, 5–31.

Arnold, D., & Blackburn, S. (2004). *Telling lives in India: Biography, autobiography and life history*. Delhi: Permanent Black.

Baillie Smith, M., & Jenkins, K. (2011). Disconnections and exclusions: Professionalization, cosmopolitanism and (global?) civil society. *Global Networks, 11*, 160–179.

Baillie Smith, M., & Jenkins, K. (2012). Existing at the interface: Indian NGO activists as strategic cosmopolitans. *Antipode, 44*, 640–662.

Baillie Smith, M., Laurie, N., Hopkins, P., & Olson, E. (2013). International volunteering, faith and subjectivity: Negotiating cosmopolitanism, citizenship and development. *Geoforum, 45*, 126–135.

Bankoff, G. (2001). Rendering the world unsafe: 'Vulnerability' as western discourse. *Disasters, 25*, 19–35.

Batliwala, S. (2002). Grassroots movements as transnational actors: Implications for global civil society. *Voluntas: International Journal of Voluntary and Non-profit Organizations, 13*, 393–409.

Behar, R. (1996). *The vulnerable observer: Anthropology that breaks your heart*. Boston, MA: Beacon Press.

Berry, C., & Gabay, C. (2009). Transnational political action and 'Global Civil Society' in practice: The case of Oxfam. *Global Networks, 9*, 1470–2266.

Burgos, E. (1994). *Me Llamo Rigoberta Menchú y Así me Nació la Conciencia* [I, Rigoberta Menchu]. Mexico City: Siglo Veintiuno.

Chacko, E. (2004). Positionality and praxis: Fieldwork experiences in rural India. *Singapore Journal of Tropical Geography, 25*, 51–63.

Chatterjee, P. (2008). Democracy and economic transformation in India. *Economic and Political Weekly, 42*, 53–62.

Cooke, B., & Kothari, U. (Eds.). (2001). *Participation: The new tyranny?* London: Zed Books.

Cotterill, P., & Letherby, G. (1993). Weaving stories: Personal auto/biographies in feminist research. *Sociology, 27*, 67–79.

Crush, J. (Ed.). (1995). *Power of development*. London: Routledge.

Datta, A. (2009). Places of everyday cosmopolitanisms: East European construction workers in London. *Environment and Planning A, 41*, 353–370.

England, K. V. L. (1994). Getting personal: Reflexivity, positionality, and feminist research. *The Professional Geographer, 46*, 80–89.

Escobar, A. (1995). *Encountering development: The making and unmaking of the Third World*. Chichester: Princeton University Press.

Findlay, A. M. (2005). Editorial: Vulnerable spatialities. *Population, Space and Place, 11*, 429–439.

Geiger, S. (1986). Women's life histories: Method and content. *Signs: Journal of Women in Culture and Society, 11*, 334–351.

Gidwani, V. K. (2006). Subaltern cosmopolitanism as politics. *Antipode, 38*, 7–21.

Gillespie, A., & Cornish, F. (2009). Intersubjectivity: Towards a dialogical analysis. *Journal for the Theory of Social Behaviour, 40*, 19–46.

Gluck, S. B., & Patai, D. (Eds.). (1991). *Women's words: The feminist practice of oral history*. London: Routledge.

Harding, J. (2006). Questioning the subject in biographical interviewing. *Sociological Research Online, 11*. Retrieved from http://www.socresonline.org.uk/11/3/harding.html

Hart, G. (2001). Development critiques in the 1990s: culs de sac and promising paths. *Progress in Human Geography, 25*, 649–658.

Held, D. (2006). *Models of democracy*. Cambridge: Polity.

Jenkins, K. (2014). *Making the extraordinary everyday: Women anti-mining activists' narratives of staying put and carrying on in Peru and Ecuador*. Paper presented at the Society for Latin American Studies Annual Conference 2014, London.

Kachru, B. B. (1986). The Indianization of English. *English Today, 2*, 31–33.

Kagitcibasi, C. (2005). Autonomy and relatedness in cultural context: Implications for self and family. *Journal of Cross-Cultural Psychology, 36*, 403–422.

Kaldor, M. (2003). *Global civil society : An answer to war*. Cambridge: Polity.

Kothari, U. (2005). From colonial administration to development studies: A post-colonial critique of the history of development studies. In U. Kothari (Ed.), *A radical history of development studies* (pp. 47–66). London: Zed Books.

Kothari, U. (2008). Global peddlers and local networks: Migrant cosmopolitanisms. *Environment and Planning D: Society and Space, 26*, 500–516.

Larner, W., & Laurie, N. (2010). Travelling technocrats, embodied knowledges: Globalising privatisation in telecoms and water. *Geoforum, 41*, 218–226.

Lewis, D. (2008). Crossing the boundaries between 'Third Sector' and state: Life-work histories from the Philippines, Bangladesh and the UK. *Third World Quarterly, 29*, 125–141.

Lewis, D., & Mosse, D. (Eds.). (2006). *Development brokers and translators: The ethnography of aid and agencies*. Bloomfield, CT: Kumarian Press.

Liamputtong, P. (2007). *Researching the vulnerable: A guide to sensitive research methods*. London: Sage.

Liamputtong, P. (2010). *Performing qualitative cross cultural research*. Cambridge: Cambridge University Press.

Lind, A. (2003). Feminist post-development thought: "Women in Development" and the gendered paradoxes of survival in Bolivia. *Women's Studies Quarterly, 31*, 227–246.

Manzo, K. (2008). Imaging humanitarianism: NGO identity and the iconography of childhood. *Antipode, 40*, 632–657.

Mawdsley, E., Townsend, J. G., Porter, G., & Oakley, P. (2002). *Knowledge, power and development agendas: NGOs North and South*. Oxford: INTRAC.

Maxey, I. (1999). Beyond boundaries? Activism, academia, reflexivity and research. *Area, 31*, 199–208.

McKinnon, K. (2007). Postdevelopment, professionalism, and the politics of participation. *Annals of the Association of American Geographers, 97*, 772–785.

McKinnon, K. (2008). Taking post-development theory to the field: Issues in development research, Northern Thailand. *Asia Pacific Viewpoint, 49*, 281–293.

Miles, M., & Crush, J. (1993). Personal narratives as interactive texts: Collecting and interpreting migrant life-histories. *Professional Geographer, 45*, 95–129.

Nairn, K., Munro, J., & Smith, A. B. (2005). A counter-narrative of a 'failed' interview. *Qualitative Research, 5*, 221–244.

Naudé, W., Santos-Paulino, A. U., & McGillivray, M. (2009). *Vulnerability in developing countries*. Helsinki: United Nations University Press.

Russell, G. M., & Kelly, N. (2002). Research as interacting dialogic processes: Implications for reflexivity. *Forum: Qualitative Social Research*, 3(3), article 18. Retrieved from http://www.qualitative-research.net/index.php/fqs/article/view/831/1806RUSSELL

Shank, S., & Nagar, R. (2013). Retelling stories, resisting dichotomies: Staging identity, marginalization and activism in Minneapolis and Sitapur. In L. Peake & M. Rieker (Eds.), *Rethinking feminist interventions into the urban* (pp. 90–107) Abingdon: Routledge.

Slim, H., & Thompson, P. (1993). *Listening for a change: Oral testimony and development*. London: Panos Publications.

Smith, C. A. (1999). Why write an expose of Rigoberta Menchu? *Latin American Perspectives, 26*, 15–28.

Smith, F. (1996). Problematising language: Limitations and possibilities in 'foreign language' research. *Area, 28*, 160–166.

Twyman, C., Morrison, J., & Sporton, D. (1999). The final fifth: Autobiography, reflexivity and interpretation in cross-cultural research. *Area, 31*, 313–325.

UNDP. (2014). *Human development report 2014: Sustaining progress: Reducing vulnerabilities and building resilience*: Author.

Viezzer, M. (1977). *'Si me permiten hablar': Testimonio de Domitila, una mujer de las minas* [If I may be allowed to speak: Testimony of Domitila, a woman from the mines]. Mexico City: Siglo Veintiuno Editores.

Wiles, J. (2011). Reflections on being a recipient of care: Vexing the concept of vulnerability. *Social and Cultural Geography, 12*, 573–588.

Yarrow, T. (2008). Life/history: Personal narratives of development amongst NGO workers and activists in Ghana. *Africa, 78*, 334–358.

Being patient, being vulnerable: exploring experiences of general practice waiting rooms through elicited drawings

Kyle Eggleton, Robin Kearns and Pat Neuwelt

ABSTRACT

For members of vulnerable populations, articulating feelings and emotions about a place to a cultural outsider may be challenging. This article describes a study using participatory visual methodology, in which participants were invited to create drawings to explore their experiences as patients in general practice waiting rooms, and interacting with receptionists. Set in New Zealand, the *Pākehā* (white) male researcher used a traditional meeting protocol (*hui* process) as a process for trust building with the participants; most were Māori, many were female, of a different age group, and had mental health vulnerabilities. As well as explaining the research, the preamble to the interviews also involved making family connections, gift giving and a mutual sharing of ideas. Elicited drawings provided the framework for conversational interviews. Colour, materials, symbols and metaphors were jointly explored through a process of shared analysis of images created. Through this process, participants reflected on, and conceptualised, ideas that addressed disempowerment, discrimination and racism. Key to the project's success was the adoption of a culturally safe approach and willingness on the part of the researcher and participants to not only trust the process, but also be patient and vulnerable – in a context in which being a vulnerable patient prevails.

Etre patient, être vulnérable: exploration des expériences d'une salle d'attente d'un cabinet médical par demande de dessins

RÉSUMÉ

Pour les membres de populations vulnérables, l'expression de sentiments et d'émotions au sujet d'un lieu à un étranger culturel peut être difficile. Cet article décrit une étude utilisant une méthodologie visuelle participative, selon laquelle les participants sont invités à créer des dessins pour explorer leurs expériences en tant que patients dans des salles d'attentes de cabinets médicaux et à interagir avec les réceptionnistes. Situé en Nouvelle-Zélande, le chercheur (blanc) *Pākehā* a utilisé un protocole de rencontre traditionnel (hui processus) comme procédé pour établir la confiance auprès des participants ; la plupart étaient des Māori, femmes pour beaucoup, d'un groupe d'âge différent, et de santé mentale vulnérable. En plus d'expliquer la recherche, le préambule à ces entretiens incluait aussi de faire des connections familiales, d'offrir un cadeau et d'échanger mutuellement des idées. L'obtention de dessins fournissait le cadre aux conversations des entretiens. La couleur, le matériau, les symboles et les métaphores éraient explorés conjointement à travers un processus de partage d'analyses des images créées. A travers ce procédé, les participants réfléchissaient et conceptualisaient des idées qui touchaient au fait d'être défavorisé, à la discrimination et au racisme. La clé du succès du projet a été l'adoption d'une approche culturellement sans risque et la volonté de la part du chercheur et des participants de non seulement se fier au processus, mais aussi d'être patient et vulnérable – dans un contexte où domine le fait d'être un patient vulnérable.

Ser paciente, ser vulnerable: la exploración de experiencias en las salas de espera de atención primaria a través de dibujos

RESUMEN

Para los miembros de poblaciones vulnerables, articular sentimientos y emociones sobre un lugar para una persona ajena a su cultura puede llegar a ser un reto. Este artículo describe un estudio utilizando la metodología participativa visual, en la que se invitó a los participantes a crear dibujos para explorar sus experiencias como pacientes en las salas de espera de atención primaria, interactuando con los recepcionistas. Situado en Nueva Zelanda, el investigador *Pākehā* (blanco) utilizó un protocolo de reunión tradicional (proceso *hui*) como un proceso de construcción de confianza con los participantes; la mayoría eran maoríes, muchos eran de sexo femenino, de diferente edad, y con dificultades de salud mental. Así como la explicación de la investigación, el preámbulo de las entrevistas también implicó hacer conexiones familiares, dar regalos, y un reparto mutuo de ideas. El pedido de dibujos proporcionó el marco para las entrevistas conversacionales. Los colores, materiales, símbolos y metáforas fueron explorados de forma conjunta a través de un proceso de análisis compartido de imágenes creadas. A través de este proceso los participantes reflexionaron sobre, y conceptualizaron, ideas que abordaron la falta de poder, la discriminación y el racismo. La clave para el éxito del proyecto fue la adopción de un enfoque culturalmente seguro y la buena voluntad por parte del investigador y de los participantes para no sólo confiar en el proceso, sino también para ser pacientes y vulnerables — en un contexto en el que prevalece ser un paciente vulnerable.

Introduction

In the spaces of health care, the vulnerability of the encountered 'other' can be starkly evident. In this paper, we illustrate an example of how relationship building and sharing of values can be embedded in research with 'vulnerable' groups. We show that through facilitation of trust and patience, there can be a co-creation of insights that potentially challenge the power and privilege embedded in sites such as places of primary health care. We build on Dyck and Kearns' (1995, p. 138) concern for an increased 'openness to the cultural contexts of our research (which) might lead us to rethink our ways of knowing in an emerging geography of health and healing'. In so doing, we build on calls for greater community participation by encouraging two-way conversations between general practices and communities (Neuwelt, 2012) and extend the reach of geographies of primary care into the waiting room, an intimate nexus of people, practice and place (Crooks & Andrews, 2009).

We begin by noting that while patients from socioeconomically disadvantaged population groups have often been coded 'vulnerable', there is a need to highlight the problematic character of this term. Representatives of the health care system tend to deem patients as 'vulnerable' by virtue of their distance from nodes of care, limited health literacy or burden of illness relative to their own or the system's resources. Coding some patients as 'vulnerable' can be based on an institutional logic, serving to release more funds or medical attention, yet we nonetheless see the term as problematic for its capacity to overlook people's strengths and resilience.

Undertaking research with 'vulnerable' populations can be fraught with ethical issues and challenges. In the New Zealand context, power and privilege remains largely in the hands of descendants of the colonisers and other *tauiwi* (newcomers; those who possess cultural similarities to the colonisers). As in white settler countries elsewhere, such as Australia, the U.S. and Canada, colonisation has had enduring effects on the health of the indigenous Māori population (Came, 2014). Indigenous populations, once healthy and vital, have been – and continue to be – rendered 'vulnerable'. Māori make up 15% of the population (Statistics New Zealand, 2015), and have established poorer health status (Curtis et al., 2010; Marie, Fergusson, & Boden, 2008) and poorer access to primary care than non-Māori (Barwick, 2000; Davis, Lay-Yee, Sinclair, & Gribben, 1997).

Our study sought to explore the experiences of reception processes of 'vulnerable' people – notably mental health consumers, youth and parents of young children. Research suggests that members of these groups may either experience difficulties in accessing general practice or problematic relationships with receptionists (Neuwelt, Kearns, & Browne, 2015). Such patients are therefore 'vulnerable' in the conventional sense of belonging to a marginalised demographic group or having a burden of illness greater than what could be addressed by the health care system's resources.

During recruitment, it became evident that the majority of participants identified as Māori, introducing a cross-cultural dynamic to the study and prompting collective reflection among the investigators. As indigenous health status and experience have come under increasing investigative scrutiny, the act of research itself has been linked to processes of colonisation (Louis, 2007). This contention has been exacerbated by situations in which, on balance, the benefits of research have accrued to the researcher and not 'the researched' (Tuhiwai Smith, 2012). The dominant sociocultural theories underpinning research processes may negate the socio-cultural reality of a 'vulnerable' population (Wilson & Neville, 2009). Hence, researchers need to think critically about the ways in which research projects are constructed and developed, how they engage with 'vulnerable' populations, and how they conduct the research. Such considerations are prerequisites to ensuring that a project is conducted in an acceptable and appropriate manner (Wilson & Neville, 2009). The concept of cultural safety, developed by Ramsden (1990), is relevant here. This approach originated in nursing education and involves both acquiring awareness of cultural understandings and recognising the potential for exacerbating discrimination through the unacknowledged attitudes held by those in service and professional roles. Cultural safety principles have been adapted for the practice of social science research and essentially involve building trust within the research relationship (Kearns & Dyck, 2015). These principles offer pointers towards designing research that avoids endorsing vulnerability and rather allows accrual of benefits to 'the researched'.

In this article, we argue for the need to pursue trust-building practices in working with so-called 'vulnerable' people in health care settings using creative methods of engagement. We illustrate this argument through outlining elements of a New Zealand-based study exploring the experiences of 'vulnerable patients' in general practice waiting rooms, and interacting with receptionists. We outline how adopting a combination of two approaches (elicited drawings and conversational interviews) encouraged relationship building and trust. In turn, this led to a lessening of the power differential through an empowerment of otherwise 'vulnerable' participants.

We contend that the potency of this study lay in the combination of these approaches and, as a result, we share Pain and Francis' (2003, p. 46) advocacy for visual approaches which 'move beyond words and texts' and 'the emancipatory potential of techniques that appear to transcend cultural, language and literacy barriers'. Through this account, we add to a growing body of literature on visual methodologies (Rose, 2007) as well as contribute to geographies of health care and marginalised populations.

Challenges and positionality

During the design of the study, we grappled with the definition of 'vulnerability' and methods of addressing power differentials between the researchers and participants. The challenge during fieldwork was in removing the gaze from the researched 'other'. As most recruited participants were Māori, and differed from the interviewer by other factors such as gender, the first author was confronted by his own whiteness and his outsider status with the people he interviewed. As a general practitioner (GP) and social science researcher, critiquing power and privilege in general practice settings required self-reflection on his own power and privilege. This critical self-reflection evoked the notion that 'whiteness' not only concerns the skin colour of the researcher, but is also implicated in the cultural practices in white settler societies that dominate and give structural advantage to non-indigenous people (Aveling, 2013). We therefore agree with Martin-McDonald and McCarthy (2008), whose view is that although privilege and power can be invisible to the white researcher, whiteness is visible to those who are not white. An ongoing process of reflecting on the dynamics of cross-cultural fieldwork is therefore critically important. While interviews with more than one researcher present might be overwhelming for participants, an ongoing conversation about participant encounters among all members of the research team can provide a context within which to consider the challenges of positionality.

The issue of being an outsider researching across a cultural 'divide' required care and consideration. For, as Kearns and Dyck (2015), contend, researchers invariably embody power by virtue of their socio-economic class, institutional affiliation, gender or ethnicity. The differences in societal power between researcher and participants shape the degree to which the researcher is coded an insider or outsider. Indeed, asserting that there are neatly packaged insider/outsider positions ignores the multiplicity and complexity of self-identity. Relationships as an insider/outsider are constantly negotiated (Parker Webster & John, 2010) and there is fluidity in this relationship as loci of commonality becomes discovered or lost (Merriam et al., 2001). These commonalities include not only ethnic identity but also values, experiences and places. The position of the researcher 'in the field', then, is important. In this study, the first author has had longstanding personal links to Māori communities, and that forms an important part of his identity. He spent his childhood in a small rural, and predominantly Māori community in New Zealand where he was taught Māori language, entered the *marae* (communal meeting place) to attend *tangi* (Māori funerals) or *hui* (formal meetings), and learned *haka* and *waiata* (traditional dance and songs). Although he attended high school and university in a large urban centre, the memories of his childhood and the values imparted have remained with him. During his medical training, he returned to the place of his childhood and worked for a period of time as a GP, during which he reconnected with his sense of belonging to the small village where he grew up. Disillusioned with the manifestations of health inequity he saw in mainstream general practice, he moved into work as a *kaupapa* Māori medical officer – a doctor working for a Māori tribal health organisation delivering primary care under a Māori model of health. It is our contention that, on this basis, the first author is not an external outsider in Banks' (1998) terms, but rather is positioned more along the continuum towards an external insider. It is also of note that other members of the research team share some of this cultural and locational empathy: the second author spent his childhood, and has undertaken a number of research projects, in the region; and the third author was a GP and is now a researcher in the same region.

Notwithstanding this positionality, and the way insider/outsider status is negotiated through relationships and developing points of commonality, some authors question whether non-indigenous researchers should become involved in research involving indigenous peoples. Indeed, one commonly held position is that non-indigenous researchers should only act as an ally with indigenous researchers and not be the primary researcher (Aveling, 2013). The underlying premise is that non-indigenous people cannot speak about and through a worldview that they are not part of. An alternative view is that non-indigenous researchers can in fact be primary researchers if the research adheres to a number of culturally safe principles (Dyck & Kearns, 1995). The foremost principle is that the research process must be collaborative and share knowledge (Louis, 2007). Through this process, a research agenda can develop that is respectful, empathetic, ethically sound (Hodge & Lester, 2006) and potentially empowering of people deemed 'vulnerable' by the health care system. The process of addressing issues of 'vulnerability' led us to the secondary aim of the study, and the focus of this paper. This aim was to develop a research approach that would lead to a building of trust and one that shows it is possible for non-indigenous researchers to undertake research with indigenous groups. In the next section, we outline definitions of 'vulnerability'. We then describe how creative research approaches, such as using a participatory visual methodology embedded within a culturally safe framework, may assist in encountering vulnerability.

Research on/with 'vulnerable' populations

Vulnerability has various meanings. Moore and Miller (1999, p. 1034) suggest that 'vulnerable' people are those who 'lack the ability to make personal life choices, to make personal decisions, to maintain independence, and to self-determine'. Meanwhile, Flaskerud and Winslow (1998, p. 69) define 'vulnerable' populations as 'social groups who have an increased relative risk or susceptibility to adverse health outcomes'. For other writers, 'vulnerability' occurs when members of a population are at higher risk and encounter greater health burdens than others (Wilson & Neville, 2009). 'Vulnerability' for these, and many writers, implies a 'distinctive precariousness' of an individual or group of people (Kipnis, 2001, p. G5).

Within population health discourses, this 'distinctive precariousness' has led to describing populations as 'vulnerable', or at risk of being 'vulnerable', based on particular characteristics such as socio-economic status, ethnicity, gender, housing and pre-existing physical and mental health conditions (Aday, 2001). Significant evidence exists linking poor health outcomes to particular demographic groups, such as those who are socio-economically deprived (Aday, 2001; Braveman, 2006; Evans & Kantrowitz, 2002) or minority or indigenous ethnicities (Curtis et al., 2010; Marie et al., 2008). Individuals from these groups are also referred to in public health discourse as 'hard to reach', implying what at times is a convergence between issues of spatial access, and incompatibility of service design (Carr, Matheson, & Tipene-Leach, 2006). Aday (2001) describes the predictors of 'vulnerability' as derived from a person's social status and a consequent differential ability to mobilise personal and political power. Within Aday's (2001) framework poor physical, psychological and social health stem from poorer quality of health care delivered to 'vulnerable' populations or a lack of access to health care. Like Aday (2001), Kilbourne, Switzer, Hyman, Crowley-Matoka, and Fine (2006) describe health disparities as arising from gaps in quality of care. In addition, Kilbourne et al. (2006) emphasise a definition of 'vulnerable' populations as those facing discrimination. This definition suggests that 'vulnerability' occurs through bias, not just based on ethnicity but also on socioeconomic or cultural differences. The health care environment can express such bias and, thus, has particular implications for populations who experience dismissal, discrimination and social exclusion in other aspects of their everyday lives (Browne et al., 2012)

In arguing against categorical definitions of 'vulnerability' Luna (2009) suggests that if political or system responses can modify a person's 'vulnerability', or if the situation that leads to 'vulnerability' changes, then vulnerability should not be a categorical label. Rather she suggests that 'vulnerability' is multi-layered and that a flexible, creative and layered approach is required to protect research subjects and minimise 'vulnerability'. Further, emphasising a relational aspect of vulnerability, she argues that modifying the relationships between people and circumstances alters the degree of 'vulnerability'. This perspective seems infused with potential for change and hope, resonating with our view on the dynamics of general practice waiting rooms and reception spaces (see Neuwelt et al., 2015) and our views on trust building as an approach to addressing 'vulnerability'.

Many approaches to reducing vulnerability in the research setting act at the level of the researcher–participant relationship. However, this relationship is inherently unequal (Hay, 2010). When a research team includes a health professional investigating a health issue, such as in this study, this may be especially true (Hofman, 2004; Smith, 2008). Seeking actively to involve marginalised groups in research enhances the chances that the benefits of research and evaluation are distributed equally among groups (Ulrich, Wallen, & Grady, 2002) and reflects the principle of social justice (Smith, 2008). Active involvement of participants, when researchers establish relationships with those who belong to the group being studied can be considered 'culturally safe' (Kearns & Dyck, 2015). In other words, culturally safe research occurs when participants are included, respected and can trust the researcher(s). Importantly, cultural safety requires a commitment to honour the rights of research participants and to 'work with' those being researched (Wilson & Neville, 2009). Cultural safety inherently attempts to deconstruct the power relationship in the researcher–participant dyad by empowering participants (Papps & Ramsden, 1996). One mechanism of handing back some control within the research encounter is to ensure that the methods used engage with non-verbal ways of knowing.

Increasingly, arts-based research is used as an approach to engage with the lived experience of 'vulnerable' or 'hard to reach' people (Gerstenblatt, 2013; Redwood, Gale, & Greenfield, 2012; Scott, 2009). Wang (1999), for example, suggested that photovoice allows the conduct of research by women and *with* women, instead of *on* women. In addition, the use of participant-derived images enables an appreciation of women's subjective experiences. More generally, Pain and Francis (2003) suggest that participatory diagramming can be an effective method in accessing 'hard to reach' people. A critical element in accessing lived experience is the degree of engagement of participants within the research. Pain and Francis (2003) argue that methods alone do not transcend the power differential between participants and researcher. There is therefore a need for ongoing conversations within a research team to reflect critically upon the character and manifestations of vulnerability as encountered in participants and experienced by the field researcher.

In summary, 'vulnerability' is not a categorical label, but rather is a multi-layered concept that alters depending on a person's situation. Within this study, participants belonged to a marginalised or 'vulnerable' group in the conventional sense with regard to their ethnicity or health status (or both). Given that 'vulnerability' is dependent on power dynamics, we prefer to see participants as structurally disempowered rather than marginalised or 'vulnerable' per se. By this, we mean participants have had an inability to make meaningful change to economic, political, social and health systems shaping their lives due to the structural characteristics of the contexts in which they were embedded. This situation arguably places the burden of addressing power relations onto the researcher and leads to further calls for developing meaningful relationships in research as well as research approaches that facilitate the development of such relationships. Within the remainder of this paper, we outline a multi-layered approach that allowed a building of trust and an examination of structural barriers in the context of exploring of the internal geography of general practice waiting rooms.

Context of the study

The context of this study is general practice waiting rooms located in the northern area of New Zealand. General practice is the cornerstone of the primary health care system in New Zealand. Its emphasis is on episodic care for individuals and is predominantly delivered in a medical centre setting. Typically, a New Zealand general practice consists of a team of general medical practitioners (GPs or family doctors), practice nurses and receptionists. The ownership structure of practices varies across the country with a range of models: GP owner-operated practices, community trust practices, *iwi* (Māori tribe) owned practices and an increasing number of corporate practices (Neuwelt & Kearns, 2009).

Achieving equitable access to general practice has proved difficult to achieve in the New Zealand setting (Jatrana & Crampton, 2009). This inequity is most pronounced for Māori (Reid & Robson, 2007). One understudied area of access is the acceptability of general practice and, in particular, the acceptability of general practice waiting rooms. These spaces are, in practical terms, a key part of the passage into general practice and are occupied by both patients and receptionists. With the waiting room regarded as a gateway into care, the receptionist has been seen as a 'gatekeeper' (Offredy, 2002). Bestowed with a role characterised by the need for multi-tasking (Neuwelt et al., 2015), the receptionist keeps watch over a space often filled with restlessly waiting patients. Sights, smells and sound invariably interplay to create a richly sensed and felt space. While a place of pause in the journey to care, waiting rooms are (perhaps paradoxically) not static, but rather are part of the internal and mobile geography of the health care system replete with movement, as patients and receptionists engage in a choreographed fashion – or as Seamon (1980) described it, a place ballet.

Different gazes tend to operate within this space. For example, there is the disciplinarian role played by the receptionist, observing and classifying patients upon entry into roles of sick/not sick (Foucault, 1977) and the Lacanian gaze returned from an anonymous 'other' as patients enter the waiting room and survey their surroundings (Krips, 2010). With respect to the former, the computer is symbolic of a tension, and divided attention, as a receptionists' gaze becomes distracted between the person to the object (Rapport, Doel, & Elwyn, 2007). Reception area surroundings are replete with symbols evident in decorations, artwork, warmth, light and distractions being emblematic of good quality of care. In contrast, regimented chairs, darkness, coldness and minimal distractions (e.g. well-thumbed and dated magazines) represent the opposite (i.e. poorer quality of care) (Arneill & Devlin, 2002). Examples of the latter style of waiting room give rise to the contention that such spaces might be interpreted as non-places (Augé, 1992/1995). In situations where they are transactional, instructional, sterile and generative of minimal social interaction, Strathmann and Hay (2009, p. 217) describe waiting rooms as 'places of desperation and impotence ... places suffused with anxiety where one can exhibit very little agency'. This scenario was contrasted with the convivial rural waiting room settings noted by Kearns (1991) in which people arrived to engage with others, despite not seeking a medical appointment. The potential for general practice reception processes to act as barriers to people in accessing general practice is reinforced when the 'experiences' of vulnerable people are explored. In order to both acknowledge yet attempt to diminish 'vulnerability', we used a participatory visual methodology, embedded in a process of trust building.

The research process

The line of questioning taken in the research process highlighted the importance of building trust. During fieldwork, the first author asked about negative experiences of general practice from making contact through to seeing a doctor. This questioning was potentially problematic as, not only was he inadvertently representative of white and male privilege, but also the privilege associated with belonging to the medical profession. On face value, it would seem unlikely that he would be able to facilitate participants discussing their experiences in any depth. However, our experience was ultimate otherwise.

After receiving institutional ethics approval, the recruitment approach involved engaging with a range of non-government organisations (NGOs) in New Zealand's most northerly city, Whangarei (district population: 77,000, 160 km north of Auckland). Of its population, 24% are Māori, compared to 15% of the national population. In 2013, it also had a lower median income ($NZD 25 300) compared to the rest of New Zealand ($NZD 28 000). Source NGOs included a mental health consumer group, a youth drop-in-centre, a support agency for seniors and a Māori social services provider.

Table 1. Participant details.

Pseudonym	Age	Gender	Ethnicity	NGO	Interview location
Te Atawhai	71	Female	Māori	HSSP	Home
Hari	74	Male	Māori	HSSP	Home
Sharon	43	Female	Pākehā	MHCG	Home
Manuka	35	Male	Māori	HSSP	Marae
Moana	26	Female	Māori	HSSP	Home
Ngaire	33	Female	Māori/Pākehā	MHCG	Home
Hepi	18	Male	Māori	YDC	Youth drop-in centre
Moengaroa	62	Female	Māori	HSSP	Work
Susie	21	Female	Māori/Pākehā/Tongan	MHCG	Home
Tiffany	59	Female	Māori	MHCG	Home
Robyn	36	Female	Māori	HSSP	Marae
Verity	66	Female	Pākehā	SAS	Support agency for seniors office
Wiremu	76	Male	Māori	HSSP	Home

Note: *Pākehā* is a Māori word, and commonly used descriptor in New Zealand, for people who are elsewhere termed *White* and of European ancestry.
Abbreviations: HSSP: Hapū (sub tribe) social services provider; MHCG: Mental health consumer group; YDC: Youth drop-in centre; SAS: Support agency for seniors.

Each NGO determined the process by which they would recruit participants. All NGOs used a snowballing technique and potential participants were given an information sheet. This described the project and informed them that drawing was part of the research method. Additionally, the first author contacted participants by telephone to ensure that they understood the project, that they had sought general practice care within the previous year (for themselves or a family member), and to make sure that they were comfortable with drawing. After clarification of the project, two potential participants declined further involvement and 13 agreed, 11 of who were Māori.

The first author interviewed each participant in a setting of her/his choice between March and July 2014 (see Table 1). Most participants chose this setting to be their own home, but some asked for an interview on their *marae* (tribal meeting place) and one, the office of the NGO that recruited her. A number had *whānau* (family) members accompany them. The first author observed *tikanga* (common cultural protocols) by removing his shoes in each setting (except the office space), bringing *kai* (food) to share, and offering a *koha* (gift) to each participant. Interviews were guided by a cultural framework, known as 'the *hui* process'. A *hui* is a Māori term for a gathering at which cultural processes occur. These processes offer a framework for engaging with Māori in a medical setting and include the following (Lacey, Huria, Beckert, Gilles, & Pitama, 2011):

(1) *Mihi* – the initial greeting.
(2) *Whakawhanaungatanga* – making a connection.
(3) *Kaupapa* – reason for the encounter.
(4) *Poroporoaki* – concluding the encounter.

A critical component of the *hui* process is *whakawhanaungatanga,* the development of relationships between individuals and an integral component of contemporary Māori culture. Bishop (1999, p. 4) describes the whānau, or family unit, as the 'location for constructing shared common understandings and meanings' and as analogous to the relationship between researcher and participant. Bishop and Glynn (1999) summarise the elements of *whakawhanaungatanga* as:

(1) The establishment of relationships.
(2) Addressing power differentials through participatory practices.
(3) Researchers understand themselves as somatically involved in the research.

From a practical perspective, *whakawhanaungatanga* involves exploring connections to the land and understanding family relationships and possibly ancestors. The reciprocal nature of this discussion is of importance with the researcher disclosing her/his own connections and family relationships. By way of example, regardless of participant ethnicity, all interviews began with both parties sharing information about where they were from, who their ancestors were, where their ancestors came from, the significant members of their family, what places had they lived in, their jobs or life experiences and what their aspirations were. This preamble contributed to a culturally safe research process (see Kearns & Dyck, 2015), leading into an interview (discussed further below).

Each interview lasted up to two hours and investigated the participant's experience in three scenarios – making an appointment in general practice, their engagement with receptionists and spending time in the waiting room. Although the ethics agreement required us to codify the proposed questions and prompts, in practice, the interviews became more of a *kōrero* (conversation) that evolved and became a dynamic process. All participants asked personal questions of the interviewer, such as enquiring about people that they might know in common or questions about the health care system. A conversation then developed, in which participants discussed their experiences, whether negative or positive.

After discussing the scenario, the first author asked participants to draw an idealised representation of that aspect of the journey. The prompt for drawing was both asked verbally and written on a card. The cards were placed in front of each participant to act as a prompt during the drawing stage, at which time materials (coloured pencils, crayons, paint, marker pens and paper) were provided. When the drawing was complete, the first author asked the participant about its meaning through open-ended questions such as 'what does this show?' Meanings were queried and individual components of the drawings discussed (e.g. choice of colour, material, location on the page and symbols). Through both the act of drawing itself and mutual discussion of finished drawings, the research process emphasised participants' perceptions and meaning-making. This process was repeated for each scenario, resulting in each participant creating three drawings. The first author also participated in the sketching process and showed his drawings to participants at a later date. In summary, we followed a process that sought to solicit perceptions and meaning-making by attempting to generate trust and relationship building. Drawings as a catalyst to conversation allowed us to engage with users of reception spaces in a manner in which they could be co-creators of knowledge.

Our approach to participatory visual methodology built on Rose's (2007) critique of the power structures and social context inherent in visual images. The participatory nature of the methodology was emphasised by a process of shared analysis of the production of the image. Both participants and the first author analysed these images by means of three broad lines of inquiry. This inquiry was facilitated by prompts posed by the first author:

- Questions about the production of the image. For example, the context of the drawing, the response of the drawer when asked to draw an image and the events preceding the drawing.

- Questions about the image. What does the image show? How is it arranged? What significance is the colour to the drawer? What do the different components of the drawing signify? What knowledges are deployed? Is this a contradictory image (from the interview data)?
- The relationship between the image and the audience is is more than one interpretation possible? Where is the viewer positioned in relation to the components of the image? What relation does this produce between the viewer and the image?

These questions illustrate the defining feature of participatory visual methodology – that of engaging and collaborating with participants in shared meaning-making through a creatively visual medium (see also Coleman & Kearns, 2015; Guillemin, 2004a,b; Mitchell, Theron, Stuart, Smith, & Campbell, 2011). Following the contentions of Berger (1972) and Rose (2007) that there are different ways of seeing, we proceeded on the basis that visuality is shaped socially and culturally . We asked what people view when they enter a waiting room and why. These prompts sought to disrupt the implicit social control imposed through locating power in the hands of those observing and hence tacitly subordinating the observed (Foucault, 1977). This recognition has two implications – first, it acknowledges the power and gaze used by the permanent inhabitants of places (in this case receptionists); and second, it recognises the power and gaze used by researchers, and addressed in part by cultural safety concerns (Kearns & Dyck, 2015).

Finding meaning in drawing and shared analysis

We suggest that a number of factors acted to enable participants in our study to share information with a researcher who was a cultural 'outsider' (in both ethnicity and professional class). These factors involved building trust by developing relationships centred on extending an ethic of care and respect towards participants in a culturally safe manner. Whereas much recent literature linking care and place have emphasised the giving, receipt and commodification of care (e.g. Cox, 2013; England & Henry, 2013; Wiles, 2011), this observation signals the need for further exploration of the ways that care can take place within the places and processes of the research encounter. We argue that care was evident in the interpersonal interactions that occurred between participants and the first author, and that care was also demonstrated by a concern to ensure that participants' stories and the meanings of their drawings were carefully explored. These contentions are based on feedback from participants that they found the drawing process deeply significant for them and helped them to understand their experiences in a different light. We also gained feedback from a representative of one of the NGOs involved in recruiting participants that the research approach demonstrated care and respect. The first author modelled respect in the research encounter by adhering to a cultural framework (the *hui* process) when engaging with Māori participants, and by consciously avoiding any judgement of participants' drawings. The sharing of information and interpretations between participants and the researcher facilitated relationship building and a mutual understanding of ideas about the reception spaces and processes under consideration.

Building relationships through trust and rapport

Although drawings may potentially diminish the barriers that 'vulnerability' creates, we believe that this can only occur within a supportive environment, based on relationship

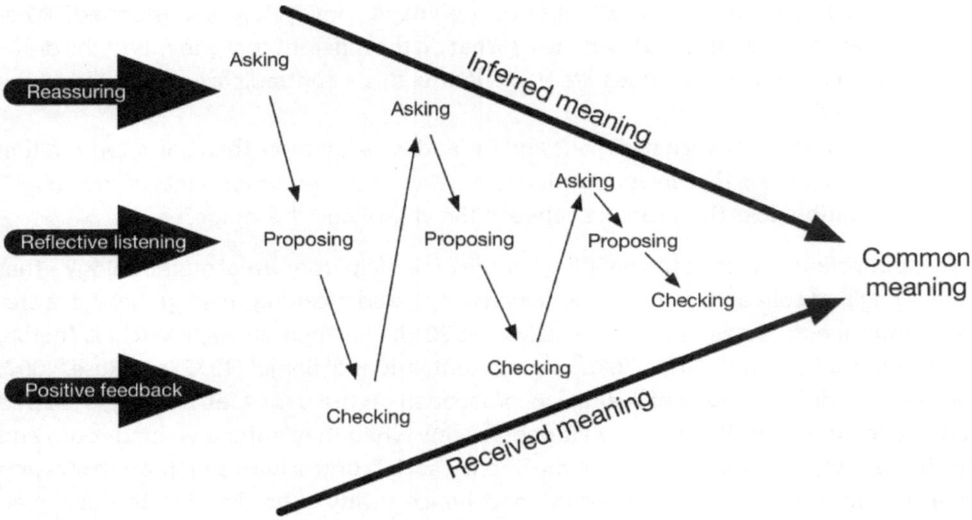

Figure 1. Reaching a common meaning.

building aided by culturally safe research practice. Meaning-making results from interactions between participants and researcher. The resulting patterns and inferences, supported by the communication style of the researcher, are part of a dynamic process. As a GP, the first author's style of communication is influenced by the consultation process developed by Neighbour (1999). This process emphasises a number of skills including rapport building, listening and eliciting skills, predictive skills and communication skills. The shared analysis in this study resembled a zig-zagging sharing of ideas, facilitated by the aforementioned skills. There was a refinement of interpretations through multiple 'readings' of the drawings and narratives until the researcher and participants reached a mutually agreed consensus. It is worth noting, however, that while some participants disagreed with occasional interpretations of their drawings, other participants may have withheld such views due to residual power differentials.

The zig-zag movement of questioning, proposing and checking meaning occurred against a background of reassurance relating to the drawings, positive feedback given to the ideas offered and a process of reflective listening. This model of narrowing the difference between inferred and received meaning is illustrated in Figure 1.

In the following dialogue, this zig-zagging communication is demonstrated, expanding on the quote pertaining to Figure 2 below:

KE: Tell me about the symbolism in the first box [Asking]. Tell me about the box itself … so you were telling me it represents you? [Reflection].

Tiffany: The green box represents me because I live by the bush.

KE: Okay so that is where the green comes in [Reflection].

Tiffany: And I chose that colour because I feel very close with nature. The environment would be my spirituality I would say so I chose that colour. And the purple and red lines inside indicates confusion and conflicting feelings.

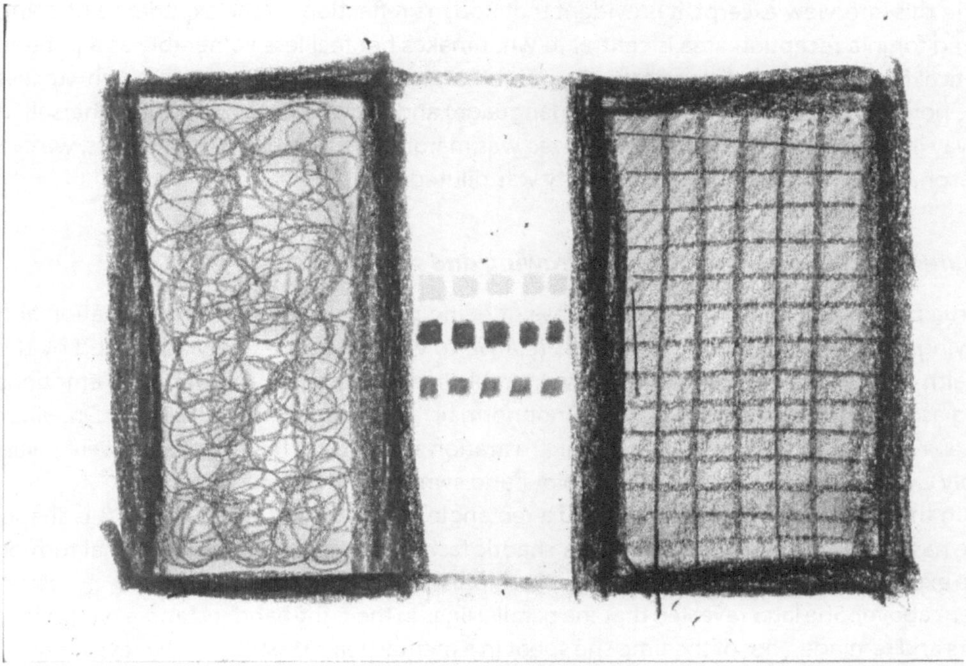

Figure 2. Feeling hurt and angry inside (by Tiffany).

KE: They are kind of like bouncing around a lot inside aren't they. They are moving around a lot. [Proposing]

Tiffany: Yeah no straight lines.

KE: A bit of turmoil. [Checking]

Tiffany: That is what it feels like to put it very simply and to draw it very simply that is what it feels like inside. And they are not happy calm colours. Red and purple, they are the more upset hurt and angry sort of colours. And they are just all over the place.

Hence, there was a jointly constructed set of meanings and metaphors identified between researcher and participant. The collaboration often occurred within a cultural context using Māori metaphors. An example lies in the following dialogue involving Robyn in which cultural metaphors and *te reo* (Māori language) develop meaning:

Robyn: Just having umm the āhua [form, figure, character] of a person when you come in is very important and that is something that you can't really explain.

KE: How does that come through in the picture?

Robyn: Well umm that is her being there and having that smile and just been easy to talk to when I come in. And having that little bit of whanaungatanga but not too much you know … Not too much to be over the top. But just that casual …

KE: It is about you as a person rather than you as a patient, it is about you as a person.

Robyn: Yeah. And I think that is what they in a perfect day they would be that little bit of caring, a little bit of empathy.

In this interview excerpt, it is evident that Robyn's reflection on the experience of being cared for in a reception area is central to what makes her feel less vulnerable as a patient. Critical for the present discussion is the point that the care applied to the research process has normalised the use of *te reo* (Maori language) and allowed Robyn to express herself in a way in which she might not have if *te reo* was marginalised. Through this process, we can reasonably assume that her vulnerability was diluted.

Shared meaning-making through drawing and shared analysis

Participants were 'vulnerable' in the sense of being part of a marginalised population and having a considerable burden of illness relative to what could be easily addressed by the health care system. Yet they were also vulnerable in another sense: they revealed emotions and aspects of themselves that would not normally be transparent to strangers. Drawings often reflected anxiety, embarrassment, frustration and anger. These emotions were invariably expressed through their use of colour and symbols as seen in Figure 2.

In this drawing, Tiffany has sketched a rectangle to symbolise herself. Inside this shape, she has drawn red and purple lines in a chaotic fashion to represent the emotional turmoil she experiences when she rings up the receptionist to make an appointment (see her statement above). She later revealed that the parallel lines in the right-hand rectangle symbolised bars and reminded her of the time she spent in a mental health institution. By revealing her emotions and personal background, Tiffany was potentially making herself more 'vulnerable' than at the outset of the research encounter. Tiffany was not the only participant to expose herself emotionally in this way. We therefore share the conclusion of other authors (Kearney, 2004; Redwood et al., 2012) that the act of drawing allows people to talk about otherwise concealed emotions and ideas.

In addition, we found that drawing not only appeared to enable people to discuss emotions that they would not normally disclose but also drawing seemed to facilitate a discussion of ideas that had not surfaced in the initial oral component of the interview. This finding resonates with that described by Mitchell et al. (2011) who found that drawings captured ideas not easily put into words. One linguistic vehicle consistently presented across research encounters was the use of metaphors, such as Tiffany's drawing of the rectangles (see Figure 2). Metaphors appeared to facilitate self-reflection or a reframing of a priori ideas (see Kearns, 1997). In acknowledgement of the 'vulnerability' of participants, both through their coded status and their self-revelation through drawing and offering personal details, the first author took great care to explore the metaphors and symbols used in drawings. Participants would often pause and look carefully at their drawings. Sometimes, they would tentatively suggest an idea and then a conversation would ensue that led to a reframing of an initial idea or a conceptualisation of a previously unspoken idea.

Ngaire, for example, demonstrated reframing (Figure 3) when she reflected on the meaning of the receptionist's desk and then decided to remove it from the drawing as it represented something that she did not agree with. Initially, she felt that desks were an integral component of being a receptionist but after drawing the desk and analysing the image, she moved from acceptance of the status quo to challenging it. This act of agency by a participant would, we contend, be less likely outside of a culturally safe process built on mutual trust.

That is clinical to me that desk, so I am taking it away (Ngaire)

Figure 3. The clinical desk (by Ngaire).

In Moana's drawing (Figure 4) she expressed something that she had not talked about in her interview: how she valued receptionists appearing similar to her.

> I suppose I drew her like that because I would like the person who was greeting my child to be comfortable so if she looks like a mother and that is relatable to her, that makes her comfortable so it just sets the whole feeling of the appointment being 'okay I am comfortable here, okay here I can do this because she looks like my mum' (Moana)

Perhaps through a combination of self-reflection, expressing emotion and creating meaning, many participants described the research process in therapeutic terms (see also Lengen, 2015). Often, this framing assisted them to understand their experiences in a different way. For example, the first author asked Sue why she drew a telephone in green when everything else in her drawing was brown. This made her realise that she had an aversion to talking by telephone and that this would have an impact on her in accessing health care unless she addressed her phone aversion.

KE: You have drawn the telephone green is there any reason or symbolism in that for you?

Sue: Yes there is. It is … I have got a thing about the phone.

KE: What does green mean?

Sue: Well I didn't want it to be red because it is not hot. It is not cold. It is not blue. It just felt … mmm You are making me really think. You know I'm going to work on this now. Because it has actually made me think about it more … I am going to have to make appointments for the rest of my life.

Participants appeared to be surprised when they discovered deeper meaning in their drawings. This discovery led to more exploratory drawings and experimentation with colour

Figure 4. Receptionists looking like Mum (by Moana).

and symbols. One participant in particular demonstrated this progression in confidence. In Manuka's initial drawing, he used pencil and tentatively sketched cartoon-like characters. His second drawing also used pencil but in addition included some symbols that represented deeper thoughts. The third drawing was rich with metaphors and colour. His increasing confidence is evident in the following quotes in response to the question 'what did you feel about the drawing'.

Drawing 1:

Manuka: umm pretty dodgy. Because I knew my drawing skill … because even drawing a square pretty … so confidence in drawing not that high.

Drawing 2:

Manuka: probably better than after the first practice and sort of thought about it a lot more and got a couple of more things written down.

Drawing 3:

Manuka: yeah quite good. Probably hit the mark better than any of the other ones of just getting a symbol and getting a bit of korero (talk) out of it.

Many of the other participants also alluded to being self-conscious about their initial drawing. Despite feeling self-conscious, almost all participants, like Manuka, developed increasing confidence with each subsequent drawing. The exceptions among the 13 participants were two elderly Māori men who expressed reluctance to draw and so were only interviewed. Other researchers have also found that people are often initially hesitant to draw (Redwood et al., 2012). Building relationships through rapport and trust prior to asking people to draw was therefore important.

Discussion: trust and vulnerability

A critical component of successful research with 'vulnerable' people lies in the building of trust between researchers and participants, and the reframing of participants as agents in the research. The importance of trust in engaging with 'vulnerable' people is well established (Emmel, Hughes, Greenhalgh, & Sales, 2007). Strategies suggested to gain initial trust include engaging with gatekeepers, as undertaken in this study (via NGOs). Other strategies include 'snowball' recruitment and immersion within a community (Crabtree & Miller, 1999). The effect of these strategies, we argue, is to transform the researcher from a stranger, embodying an institution, to a separately embodied individual. In the case of this project, the first author was affiliated with a large university in Auckland, 160 km to the south. In his engagement with gatekeepers, he positioned himself as part of the local (Whangarei) health care community. His identity as a GP was known to all the gatekeepers and this familiarity potentially aided in recruitment of participants. Emmel et al. (2007) suggest that there is another facet to the initial building of trust and that is the level of pre-existing trust between gatekeeper and potential participants. Gatekeepers with low levels of trust may confer distrust, by association, to researchers. We therefore suggest that trust has an institutional layer characterised by relationships between participants and institutions (both academic and community based).

We have built our work on a view that 'vulnerability' is inherently relational; a label derived from pathogenic or dysfunctional relationships (Lange, Rogers, & Dodds, 2013; Luna, 2009). As such, vulnerability was addressed in positive research steps focusing on relationality. As discussed earlier, a number of research approaches prioritise relationship building as a strategy for engaging with 'vulnerable' people, such as culturally safe approaches (Kearns & Dyck, 2015; Wilson & Neville, 2009), indigenous approaches (Louis, 2007; Tipa & Panelli, 2009) and participatory approaches (Pain & Francis, 2003). The relationship building in these approaches is generally interpersonal, with trust developing through partnerships, collaboration, an incorporation of cultural processes and enabling participant voice. Creative arts-based methods, as adapted for use in this study, are increasingly recognised as enabling deeper relationships to develop between researchers and 'vulnerable' people (Leavy, 2015).

Creative arts-based methods, such as participatory visual methodology, work on at least three levels. In the first instance, they can facilitate access to memories or subconscious emotions which, when depicted, may powerfully evoke a person's experience (Mitchell et al., 2011). Another level of engagement is the ability to simplify issues. Drawing in particular enables people to reduce their experiences into symbols or metaphors that can assist people in describing complex concepts (Guillemin, 2004a, 2004b). Being able to symbolically represent these thoughts may assist someone in clarifying and simplifying their intended meaning. Finally, drawings may help people in 'speaking their mind'. Some researchers have commented that they have observed participants becoming relatively disinhibited during the drawing process, sharing ideas that they would not have done as part of a conventional interview or focus group (Kearney, 2004; Redwood et al., 2012). This point is important considering the potential power dynamics inherent between 'vulnerable' participants and 'outsider' researcher.

As a medium for visual methods, the fact that the word 'drawing' is both a noun and a verb has a degree of symbolism because, as Rose (2007) notes, it is not just the product that requires analysis but also the process of production itself. Indeed, we contend that the drawing process can 'even out' power issues by providing a platform by which researcher and participants can develop deeper meaning and shared understandings. This point is of relevance when considering issues of 'vulnerability' and 'marginalisation'. People in such groups may have lower literacy levels and consequently fewer verbal skills than those of researchers. Members of 'marginalised' groups, therefore, may be less likely to communicate their ideas and concerns if researchers use a purely verbal method of data collection. In addition, discourse analysis would suggest that language is not a completely trustworthy way of communicating information for what someone says takes on a meaning that is dependent on the context of the listener (Nairn & McCreanor, 1991). In other words, oral communication runs the risk of missing meaning. Visual methodologies can be a potent alternative and contribute to an ethic of cultural safety when there are social and cultural differences between researcher and participants and offer less likelihood of a mismatch between intended and received meaning.

Visual methodologies have been criticised for placing undue expectations on a person to create an object of art (Mason, 2002). Some people may feel pressured, that their artistic skills are inadequate, or that they will be criticised. Two of the participants in this study felt uncomfortable in drawing, which suggests that visual methodologies may paradoxically increase the vulnerability of some participants. Just as we have argued that some people are not able to express themselves verbally, but may be able to do so visually, the converse is also true. We therefore suggest that in the recruitment of participants, careful attention should be paid to ensuring that participants are aware that drawings are part of the research method. Drawing on *both* drawings *and* talk is therefore important as is not referring to the images created as 'art' (given that word's implications of creative expertise and sophistication).

We have demonstrated a range of interpersonal relationship building strategies. We have shown that through *whakawhanaungatanga*, initial engagement moves the relationship from an institutional relationship to a personal one. Further, we have demonstrated that listening and communication skills embedded within a participatory and collaborative method of drawing lead to deeper engagement and meaning-making. We therefore contextualise a second layer of trust built upon interpersonal relationships and encapsulating engagement and meaning-making. In this, we contend that appropriate and culturally safe methodologies are critical to developing interpersonal relationships. We also agree with Pain and Francis (2003) that methods alone do not, in themselves, lead to trust.

Trust in oneself stems from an intrapersonal relationship and a central dynamic is the issue of power. As Pain and Francis (2003) suggest, transcending power relations is necessary in order to undertake ethical research with 'vulnerable' people. Empowerment, through 'power-within' is evident when people develop confidence and competency (Hur, 2006). Pease (2002) suggests that developing alternative knowledges, through participants providing an interpretive framework to their experiences, generates empowerment. Within our study, participants developed increasing confidence in their drawings. In addition, they provided their own interpretation that changed as they spent time reflecting, suggesting that a process of empowerment occurred for them. While empowerment reflects enhanced self-efficacy of 'vulnerable' participants, it does not have the effect of diminishing the power of the researcher. Yet, it has been argued that a dilution of the power of researchers and institutions is important in order to address 'vulnerability' (Fineman, 2008; Pain & Francis, 2003). Critical self-reflection and careful positioning may help diminish researcher power (Merriam et al., 2001).

The focus, in this paper, has been on addressing the 'vulnerability' of participants. Within the study, the first author attempted to become more vulnerable by sharing his own drawings with participants as part of the process of disseminating the findings, as well as sharing personal information about himself as part of the *hui* process. However, no data were collected regarding whether participants felt less 'vulnerable' as a result of this research process. We therefore suggest that had we embedded this participatory visual methodology within a self-consciously participatory action research paradigm a more obvious muting of power differentials may have resulted.

The final limitation of this study is that despite our attempts to engage in a culturally safe manner with the predominately Māori participants, potentially more revealing and empowering results may have resulted, if the study had been undertaken utilising a Kaupapa Māori approach (Tuhiwai Smith, 2012). Kaupapa Māori research involves research for Māori, by Māori using a Māori world view (Tuhiwai Smith, 2012). The conceptualisation of the study, the analysis of the data and the dissemination of the results would likely take on a slightly but significantly different orientation in a Kaupapa Māori study.

Conclusion

We have argued that engaging with 'vulnerable' people in a primary health care setting for research purposes is best undertaken indirectly through creative approaches such as using interactive visual methodologies. A critical element of this study that warrants adaptation elsewhere was the use of an appropriate cultural framework, such as the *hui* process, that enabled use of appropriate metaphors and the building of a culturally safe and trusting environment. For participants, the *hui* process provided a platform that diminished the power between themselves and the first author by diminishing the researcher's strangeness. This process also provided a platform upon which trust could develop, facilitated by a variety of communication skills that enabled an iterative, visually based exploration of waiting room spaces and processes. The fruits of this exploration, among otherwise 'marginalised' patients, was a rich data-set that allowed two layers of interpretation – initially in partnership, and then on the part of the research team in light of observations in the literature.

We have shown a route to a meaningful participatory strategy (participants as co-creators of knowledge and empowerment through research process). A key part of this route to greater empathy as well as problematising of vulnerability has been our adoption and use of Māori concepts. While their use in this article is foreign to some international readers, their use within our (non-Māori) research team has enhanced our ability to think, and write, our way through processes nuanced by local culture and tradition. It has allowed us to be open to being changed through the research process. Hence, to be culturally safe in research is to enter a partnership with another person or members of a population group and allow them to participate in co-creating a deeper understanding of their world (Kearns & Dyck, 2015). We have illustrated the application of these principles to an example of research practice – one in which the contours of conduct are more clearly evident given the bicultural context. We have argued that the term 'vulnerability' can be reclaimed from largely being applied *to* groups. Through the act of acknowledging structural disempowerment in the spaces of health care, we render ourselves vulnerable through the mutual exploration of feelings that elicited drawing approaches allow. The quest for mutual trust is imperative.

To practice, *whakawhanaungatanga* (acknowledgement of connectedness) potentially allows both the researcher and the 'researched' to appreciate a more fully human encounter than is often the case in research. Consciously trying to build trust and both metaphorically and literally drawing on experiences of the processes of accessing primary care, an understanding of the contours of place was sketched. By stretching senses beyond the spoken word, and trusting the process of encounter, new knowledge at the margins of health geography emerges.

Disclosure statement

No potential conflict of interest was reported by the authors.

Funding

This work was supported by the Health Research Council of New Zealand [grant number 13/753] and Auckland Faculty Research and Charitable Trust, Royal New Zealand College of General Practitioners.

ORCID

Kyle Eggleton http://orcid.org/0000-0001-5645-8326

References

Aday, L. A. (2001). *At risk in America: The health and health care needs of vulnerable populations in the United States* (2nd ed.). San Francisco, CA: Jossey-Bass.

Arneill, A. B., & Devlin, A. S. (2002). Perceived quality of care: The influence of the waiting room environment. *Journal of Environmental Psychology, 22*, 345–360. doi:10.1006/jevp.2002.0274

Augé, M. (1992/1995). *Non-places. Introduction to an anthropology of supermodernity*. London: Verso.

Aveling, N. (2013). 'Don't talk about what you don't know': On (not) conducting research with/in indigenous contexts. *Critical Studies in Education, 54*, 203–214. doi:10.1080/17508487.2012.724021

Banks, J. A. (1998). The lives and values of researchers: Implications for educating citizens in a multicultural society. *Educational Researcher*, 4–17.

Barwick, H. (2000). *Improving access to primary care for Maori and Pacific peoples*. Wellington: Health Funding Authority.

Berger, J. (1972). *Ways of seeing*. London: Penguin.

Bishop, R. (1999). Kaupapa Māori research: An indigenous approach to creating knowledge. In N. Robertson (Ed.), *Māori and psychology: Research and practice* (pp. 1–6). Hamilton: Māori and Psychology Research Unit.

Bishop, R., & Glynn, T. (1999). Researching in Maori contexts: An interpretation of participatory consciousness. *Journal of Intercultural Studies, 20*, 167–182. doi:10.1080/07256868.1999.9963478

Braveman, P. (2006). Health disparities and health equity: Concepts and measurement. *Annual Review of Public Health, 27*, 167–194. doi:10.1146/annurev.publhealth.27.021405.102103

Browne, A., Varcoe, C., Wong, S., Smye, V., Lavoie, J., Littlejohn, D., & Lennox, S. (2012). Closing the health equity gap: Evidence-based strategies for primary health care organizations. *International Journal for Equity in Health, 11*, 59.

Came, H. (2014). Sites of institutional racism in public health policy making in New Zealand. *Social Science and Medicine, 106*, 214–220.

Carr, S., Matheson, D., & Tipene-Leach, D. (2006). The public health response to 'hard to reach populations'. In D. Pencheon, C. Guest, D. Melzer, & J. A. Muir Gray (Eds.), *Oxford handbook of public health practice* (2nd ed., pp. 266–275). Oxford: Oxford University Press.

Coleman, T. M., & Kearns, R. A. (2015). The role of bluespaces in experiencing place, aging and wellbeing: Insights from Waiheke Island, New Zealand. *Health & Place, 35*, 206–217.

Cox, R. (2013). Gendered spaces of commoditised care. *Social & Cultural Geography, 14*, 491–499.

Crabtree, B. F., & Miller, W. L. (1999). *Doing qualitative research* (2nd ed.). Thousand Oaks, CA: Sage.

Crooks, V. A., & Andrews, G. J. (Eds.). (2009). *Primary health care: People, practice, place*. Aldershot: Ashgate Press.

Curtis, E., Harwood, M., Riddell, T., Robson, B., Harris, R., Mills, C., & Reid, P. (2010). Access and society as determinants of ischaemic heart disease in indigenous populations. *Heart, Lung and Circulation, 19*, 316–324. doi:10.1016/j.hlc.2010.04.129

Davis, P., Lay-Yee, R., Sinclair, O., & Gribben, B. (1997). Maori/non-Maori patterns of contact, expressed morbidity and resource use in general practice: Data from the Waikato medical care survey 1991–2. *NZMJ, 110*, 390–392.

Dyck, I., & Kearns, R. A. (1995). Transforming the relations of research: Towards culturally safe geographies of health and healing. *Health and Place, 1*, 137–147.

Emmel, N., Hughes, K., Greenhalgh, J., & Sales, A. (2007). Accessing socially excluded people – Trust and the gatekeeper in the researcher–participant relationship. *Sociological Research Online, 12*(2), doi:10.5153/sro.1512

England, K., & Henry, C. (2013). Care work, migration and citizenship: International nurses in the UK. *Social & Cultural Geography, 14*, 558–574. doi:10.1080/14649365.2013.786789

Evans, G. W., & Kantrowitz, E. (2002). Socioeconomic status and health: The Potential role of environmental risk exposure. *Annual Review of Public Health, 23*, 303–331. doi:10.1146/annurev.publhealth.23.112001.112349

Fineman, M. A. (2008). The vulnerable subject: Anchoring equality in the human condition. *Yale Journal of Law & Feminism, 20*, 8–40.

Flaskerud, J. H., & Winslow, B. J. (1998). Conceptualizing vulnerable populations health-related research. *Nursing Research, 47*, 69–78.

Foucault, M. (1977). *Discipline & punish: The birth of the prison*. New York, NY: Pantheon Books.

Gerstenblatt, P. (2013). Collage portraits as a method of analysis in qualitative research. *International Journal of Qualitative Methods, 12*, 294–309.

Guillemin, M. (2004a). Embodying heart disease through drawings. *Health, 8*, 223–239. doi:10.1177/1363459304041071

Guillemin, M. (2004b). Understanding illness: Using drawings as a research method. *Qualitative Health Research, 14*, 272–289. doi:10.1177/1049732303260445

Hay, I. (Ed.). (2010). *Qualitative research methods in human geography* (3rd ed.). Don Mills: Oxford University Press.

Hodge, P., & Lester, J. (2006). Indigenous research: Whose priority? Journeys and possibilities of cross-cultural research in geography. *Geographical Research, 44*, 41–51. doi:10.1111/j.1745-5871.2006.00370.x

Hofman, N. G. (2004). Toward critical research ethics: Transforming ethical conduct in qualitative health care research. *Health Care for Women International, 25*, 647–662.

Hur, M. H. (2006). Empowerment in terms of theoretical perspectives: Exploring a typology of the process and components across disciplines. *Journal of Community Psychology, 34*, 523–540. doi:10.1002/jcop.20113

Jatrana, S., & Crampton, P. (2009). Primary health care in New Zealand: Who has access? *Health Policy, 93*(1), 1–10. doi:10.1016/j.healthpol.2009.05.006

Kearney, S. (2004). Drawing out emotions: The use of participant-produced drawings in qualitative inquiry. *Qualitative Research, 4*, 361–382. doi:10.1177/1468794104047234

Kearns, R., & Dyck, I. (2015). Being a culturally safe researcher. In D. Wepa (Ed.), *Cultural safety in Aotearoa New Zealand* (pp. 125–138). Auckland: Pearson Publications.

Kearns, R., & Neuwelt, P. M. (2009). Within and beyond clinics: Primary health care and community participation. In V. A. Crooks & G. J. Andrews (Eds.), *Primary health care : People, practice, place* (pp. 203–220). Aldershot: Ashgate Press.

Kearns, R. A. (1991). The place of health in the health of place: The case of the Hokianga special medical area. *Social Science & Medicine, 33*, 519–530.

Kearns, R. A. (1997). Narrative and metaphor in health geographies. *Progress in Human Geography, 21*, 269–277.

Kilbourne, A. M., Switzer, G., Hyman, K., Crowley-Matoka, M., & Fine, M. J. (2006). Advancing health disparities research within the health care system: A conceptual framework. *American Journal of Public Health, 96*, 2113–2121. doi:10.2105/AJPH.2005.077628

Kipnis, K. (2001). Vulnerability in research subjects: A bioethical taxonomy. In *Ethical and policy issues in research subjects involving human participants* (pp. G1–G13). Bethseda, MD: National Bioethics Advisory Commission.

Krips, H. (2010). The politics of the gaze: Foucault, Lacan and Žižek. *Culture Unbound: Journal of Current Cultural Research, 2*, 91–102.

Lacey, C., Huria, T., Beckert, L., Gilles, M., & Pitama, S. (2011). The hui process: A framework to enhance the doctor–patient relationship with Māori. *NZMJ, 124*, 72–78.

Lange, M. M., Rogers, W., & Dodds, S. (2013). Vulnerability in research ethics: A way forward. *Bioethics, 27*, 333–340. doi:10.1111/bioe.12032

Leavy, P. (2015). *Method meets art: Arts based research practice* (2nd ed.). New York, NY: The Guilford Press.

Lengen, C. (2015). The effects of colours, shapes and boundaries of landscapes on perception, emotion and mentalising processes promoting health and well-being. *Health Place, 35*, 166–177. doi:10.1016/j.healthplace.2015.05.016

Louis, R. P. (2007). Can you hear us now? Voices from the margin: Using indigenous methodologies in geographic research. *Geographical Research, 45*, 130–139. doi:10.1111/j.1745-5871.2007.00443.x

Luna, F. (2009). Elucidating the concept of vulnerability: Layers not labels. *International Journal of Feminist Approaches to Bioethics, 2*, 121–139.

Marie, D., Fergusson, D. M., & Boden, J. M. (2008). Ethnic identification, social disadvantage, and mental health in adolescence/young adulthood: Results of a 25 year longitudinal study. *Australian and New Zealand Journal of Psychiatry, 42*, 293–300. doi:10.1080/00048670701787644

Martin-McDonald, K., & McCarthy, A. (2008). 'Marking' the white terrain in indigenous health research: Literature review. *Journal of Advanced Nursing, 61*, 126–133. doi:10.1111/j.1365-2648.2007.04438.x

Mason, J. (2002). *Qualitative researching* (2nd ed.). London: Sage.

Merriam, S. B., Johnson-Bailey, J., Lee, M., Kee, Y., Ntseane, G., & Muhamad, M. (2001). Power and positionality: Negotiating insider/outsider status within and across cultures. *International Journal of Lifelong Education, 20*, 405–416. doi:10.1080/02601370120490

Mitchell, C., Theron, L., Stuart, J., Smith, A., & Campbell, Z. (2011). Drawings as research method. In L. Theron, C. Mitchell, A. Smith, & J. Stuart (Eds.), *Picturing research: Drawing as visual methodology* (pp. 3–10). Rotterdam: Sense.

Moore, L. W., & Miller, M. (1999). Initiating research with doubly vulnerable populations. *Journal of Advanced Nursing, 30*(5), 1034–1040.

Nairn, R. G., & McCreanor, T. N. (1991). Race talk and common sense: Patterns in Pākehā discourse on Maori/Pākehā relations in New Zealand. *Journal of Language and Social Psychology, 10*, 245–262.

Neighbour, R. (1999). *The inner consultation*. Reading: LibraPharma.

Neuwelt, P. (2012). Community participation in primary care: What does it mean 'in practice'? *Journal of Primary Health Care, 4*, 30–38.

Neuwelt, P. M., Kearns, R. A., & Browne, A. J. (2015). The place of receptionists in access to primary care: Challenges in the space between community and consultation. *Social Science & Medicine, 133*, 287–295. doi:10.1016/j.socscimed.2014.10.010

Offredy, M. (2002). Access to primary care: Decision making by GP receptionists. *British Journal of Community Nursing, 7*, 480–485.

Pain, R., & Francis, P. (2003). Reflections on participatory research. *Area, 35*, 46–54.

Papps, E., & Ramsden, I. (1996). Cultural safety in nursing: The New Zealand experience. *International Journal for Quality in Health Care, 8*, 491–497.

Parker Webster, J., & John, T. A. (2010). Preserving a space for cross-cultural collaborations: An account of insider/outsider issues. *Ethnography and Education, 5*, 175–191. doi:10.1080/17457823.2010.493404

Pease, B. (2002). Rethinking empowerment: A postmodern reappraisal for emancipatory practice. *British Journal of Social Work, 32*, 135–147.

Ramsden, I. (1990). *Kawa whakaruruhau: Cultural safety in nursing education in Aotearoa*. Wellington: Ministry of Education.

Rapport, F., Doel, M. A., & Elwyn, G. (2007). Snapshots and snippets: General practitioners' reflections on professional space. *Health & Place, 13*, 532–544. doi:10.1016/j.healthplace.2006.07.005

Redwood, S., Gale, N. K., & Greenfield, S. (2012). 'You give us rangoli, we give you talk': Using an art-based activity to elicit data from a seldom heard group. *BMC Medical Research Methodology, 12*, 7. doi:10.1186/1471-2288-12-7

Reid, P., & Robson, B. (2007). Understanding health inequities. In B. Robson & R. Harris (Eds.), *Hauora: Maori standards of health IV. A study of the years 2000–2005* (pp. 3–10). Wellington: Te Rōpū Rangahau Hauora a Eru Pōmare.

Rose, G. (2007). *Visual methodologies: An introduction to the interpretation of visual materials*. London: Sage.

Seamon, D. (1980). Body-subject, time-space routines, and place-ballets. In A. Buttimer & D. Seamon (Eds.), *The human experience of space and place* (pp. 148–165). London: Croom Helm.

Scott, A. (2009). Illness meanings of AIDS among women with HIV: Merging immunology and life experience. *Qualitative Health Research, 19*, 454–465. doi:10.1177/1049732309332707

Smith, L. J. (2008). How ethical is ethical research? Recruiting marginalized, vulnerable groups into health services research. *Journal of Advanced Nursing, 62*, 248–257.

Statistics New Zealand. (2015). *How is our population changing?* Retrieved from http://www.stats.govt.nz/browse_for_stats/people_and_communities/maori/maori-population-article-2015

Strathmann, C. M., & Hay, M. C. (2009). Working the waiting room: Managing fear, hope, and rage at the clinic gate. *Medical Anthropology, 28*, 212–234. doi:10.1080/01459740903070840

Tipa, G., & Panelli, R. (2009). Beyond 'someone else's agenda': An example of indigenous/academic research collaboration. *New Zealand Geographer, 65*, 95–106. doi:10.1111/j.1745-7939.2009.01152.x

Tuhiwai Smith, L. (2012). *Decolonizing methodologies: Research and indigenous peoples* (2nd ed.). London: Zed Books.

Ulrich, C., Wallen, G., & Grady, C. (2002). Research vulnerability and patient advocacy: Balance-seeking perspectives for the clinical nurse scientist? *Nursing Research, 51*, 71–72.

Wang, C. C. (1999). Photovoice: A participatory action research strategy applied to women's health. *Journal of Women's Health, 8*, 185–192.

Wiles, J. (2011). Reflections on being a recipient of care: Vexing the concept of vulnerability. *Social & Cultural Geography, 12*, 573–588. doi:10.1080/14649365.2011.601237

Wilson, D., & Neville, S. (2009). Culturally safe research with vulnerable populations. *Contemporary Nurse, 33*, 69–79.

The ethnographic novel as activist mode of existence: translating the field with homeless people and beyond

Michele Lancione

ABSTRACT

The paper argues in favour of creative methodologies as tool for relevant academic praxis. It provides the analysis of a concrete case in which a non-academic text – a composite book made of a participant's introduction, an ethnographic novel, 21 graphic illustrations and a political essay – allowed for a meaningful re-appropriation of a fieldwork with homeless people in Italy. Such re-appropriation is understood and theorized as a research-activist 'mode of existence', namely as a way to use creative methodologies to pursue active and emancipatory engagement with vulnerable groups. The paper analyses this 'mode of existence' as a process made of several 'translations', or orientations towards the same interests, and it clearly shows the role of non-human agencies in their unfolding. Through its more-than-human narrative, the paper provides an innovative contribution to debates on research-activism and a new reflection on how to engage meaningfully with vulnerable groups. The conclusion highlights areas of improvement to further strengthen the activist-research mode of existence presented in the paper.

Le roman ethnographique comme mode d'existence activiste: traduction du domaine des sans-abris et au-delà

RÉSUMÉ

Cet article défend les méthodologies créatives en tant qu'outil pour les pratiques universitaires pertinentes. Il fournit l'analyse d'un cas concret où un texte non-universitaire – un livre d'assemblages comprenant l'introduction d'un participant, un roman ethnographique, 21 illustrations graphiques et un essai politique – a permis la réappropriation importante d'un travail sur le terrain avec des personnes sans-abri en Italie. Une telle réappropriation est comprise et théorisée en tant que « mode d'existence » de recherche activiste, c'est-à-dire comme une façon d'utiliser les méthodologies créatives pour poursuivre un engagement actif et émancipatoire auprès des groupes vulnérables. L'article analyse ce « mode d'existence » comme un processus fait de plusieurs « traductions », ou une orientation vers les mêmes intérêts et il démontre clairement le rôle des facteurs non-humains dans leur déroulement. A travers son récit plus-qu'humain, cet article fournit une contribution innovatrice aux débats sur la recherche de l'activisme et une nouvelle réflexion sur la façon de communiquer de manière significative avec les groupes vulnérables. La conclusion met en avant les domaines d'amélioration pour renforcer encore davantage le mode d'existence de la recherche activiste présenté dans cet article.

La novela etnográfica como modo activista de existencia: traduciendo el campo con personas sin hogar y más allá

RESUMEN
En el documento se argumenta a favor de metodologías creativas como herramientas para la praxis académica relevante. Se proporciona el análisis de un caso concreto en el que un texto no académico – un libro compuesto de la introducción de un participante, una novela etnográfica, 21 ilustraciones gráficas y un ensayo político – permitió una significativa re-apropiación de un trabajo de campo con personas sin hogar en Italia. Tal re-apropiación se entiende y se teoriza como un 'modo de existencia' activista en la investigación, es decir, como una manera de utilizar metodologías creativas para buscar un compromiso activo y emancipador con grupos vulnerables. El documento analiza este 'modo de existencia' como un proceso compuesto de varias 'traducciones', u orientación hacia los mismos intereses, y muestra claramente el papel de los organismos no humanos en su despliegue. A través de su narrativa más que humana, el documento proporciona una contribución innovadora a los debates sobre el activismo en la investigación y una nueva reflexión sobre la forma de participar de manera significativa con grupos vulnerables. La conclusión destaca las áreas de mejora para fortalecer aún más el modo de existencia activista en la investigación presentado en el documento.

Prologue

September 2009, Turin, Italy: That afternoon I had an appointment with Antonio, a middle-aged homeless man who I met a few days earlier in the soup kitchen where I was volunteering. Sitting at a bar in the city centre, stirring our caffè corretto, Antonio was the one to bring up the subject. I still remember his first question very clearly: 'What do you want to know? What do you want from me?' Back then, those words sounded like a challenge. Although I replied – giving him an overview of my research intent – I couldn't stop thinking about that question: 'Che cosa vuoi da me?' (What do you want from me?). Towards the end of our conversation, Antonio brought up the subject once again. 'So, what are you going to do with this material?' This was another challenging question: my work was going to be written in English (since my PhD was based in Durham, UK), to be defended and presented far from the places he knew, the city where he was performing, the institutions with which he dealt. He listened to me attentively, while I tried to offer an overview of the above. When I concluded, he asked if I thought Turin's social services would read my work. I replied that I did not know for sure, but that I wanted to do my best to let them know about my findings.

Introduction

That first conversation with Antonio – who became a friend and an important gatekeeper for me – sowed the seeds of doubt in my mind. Was it enough to do a 'solid' piece of research? Was it sufficient to do 'my best' to communicate its outcomes to the policy-makers in Turin?

At the time, I was still uncertain about what I had to do. But the more I delved into the field, the more I felt I had to do something: I needed to 'bring' my research's outcomes closer to homeless people, policy-makers and the general public. That urgent feeling eventually took the form of a book called 'Il numero 1', which was published in 2011 by an independent publisher in Turin and distributed nationwide. Since one of the book's aims was to translate the field research in order to connect it to different audiences, 'Il numero 1' is a composite book. It includes a preface written by a homeless friend, a full-length ethnographic novel, 20 black and white illustrations, and a more academically sound appendix.

Exploring the making and the implications of the book, the aim of this paper is to offer geographers and other scholars using ethnographic methods a theoretical, empirical and critical reflection on the role creative methodologies can play in producing new temporary alliances for social change. The literature already offers examples in this sense, such as the recent collection of Marston and De Leeuw's (2013) and Eshun and Madge's (2012) post-colonial engagement with poetry as a mean to explore, discuss and communicate research findings, urban ethnologists experimenting with rhythms and flows to (non)represent the city (Atooh, 2011) or the use of videography and the tools offered by Web 2.0 to make sense of post-revolutionary spaces (Governa & Puttilli, 2016) and social conflict (Sandercock & Attili, 2012). Building on these and other interventions, the novel contribution of this paper can be summarized in three points.

First, the paper focuses specifically on the relationships between long-term ethnographic fieldwork, creative writing and vulnerable groups (a terrain that has been only partially charted in Geography, most notably by Christensen, 2012). Second, it frames these relationships in terms of the engagement and relevance that a researcher may have in her/his chosen field and context, hence, contributing to a long-standing and growing debate on activism and research (The Autonomous Geographies Collective, 2010) and on the commitments of researchers towards vulnerable groups (Cloke, Cooke, Cursons, Milbourne, & Widdowfield, 2000). Third, the paper analyses how those alliances are assembled together, and it questions the role of non-human agencies in their making. The aim is to excavate how the peculiar research-activist *mode of existence* that characterizes the book unfolded and came to the fore (Latour, 2013). How, in other words, did a particular modality of research dissemination and collaboration such as 'Il numero 1' become actualized? What are the affiliations that one could deem as generative of the short- and long-term implications of the work of the book? What can be learned from this case in order to write and do more progressive and collaborative accounting of life at the margins? The paper confronts these questions, employing the notion of *translation*, which is here understood as the movement aligning particular actants towards the fulfilment of a specific mode-of-existence (Callon, 1986; Latour, 2005; Law, 2008).

In order to ground what follows, it is worth recalling why I felt the urgency to embark on this project. There are two interrelated reasons. The first is that, at the time of my fieldwork, I was faced with a linguistic problem, which could be understood also as a problem of *space*. As Julia Christensen has summarized in her contribution to storytelling and participatory research, there is a 'significant disconnection between the emotive, personal stories of homelessness that I was collecting' and 'more conventional, and often less emotive, approaches to research dissemination' (Christensen, 2012, p. 232). What Christensen describes is a non-relation: A mismatch that does not allow communication to be expressed and meaning to be (co)constructed. Such disconnection becomes a burden upon the capacity of individuals to cooperate, to work together, to share emotions, to imagine an alternative future and, above all, to share a common space. Through academic language, I wasn't able to challenge marginalization, but I was concretely (re)enacting it (Mezzadra & Neilson, 2013): People were not able to understand, to join in, to make use of what I was saying and, in the end, a space of encounter was not generated.

Second, I was faced with a *temporal* problem. As Pickerill (2014, p. 25) has recently pointed out, vulnerable groups and people working with them 'often request immediate support, immediate access to research results, and immediate impact'. This is, of course, at odds with the time necessary to do and to disseminate academic research. To put it simply, by the time my second paper on homelessness in Turin came out (Lancione, 2014), one of my research collaborators had already died, while the City administration faced new issues, including the refugee 'crisis'. In short, if I had relied only on that and subsequent publications, I would not have been able to respond to the most immediate needs and expectations of my research collaborators. If many homeless people saw the use of telling a broader story that lasted beyond them – one able to change policies and interventions in the long run – what all of them wanted was to hear their stories told aloud in the here and now.

It is at the intersection of this *space of relations* and *time of immediacy* that the ethics and politics fuelling 'Il numero 1' emerged. Space and time were demanding something, and doing 'Il numero 1' was just a way to take that call seriously. What space and time demanded was a twofold *politics of representation* (Lassiter, 2001): One about collaboration, engagement and empowerment, in the short-run, and about a longer term self-sustaining legacy. *A posteriori*, I can confidently say that the politics fuelling 'Il numero 1' was both 'about servicing and informing powerful groups and making policies work better' – inasmuch as it was aimed at service providers in Turin – '[and] about representing and empowering marginalized people' (Pain, 2003, p. 651) – inasmuch as it was aimed at making them part of their own representation. Doing the book was about taking the contingencies of the fieldwork seriously and acting upon (and through) them.

The paper is organized as follows. In the next section, I present the three main tenets – and literatures – around which the argument is built and to which it contributes: activism, non-human agencies and ethno-fictive writing. Then, the paper presents the different translations through which the research-activist mode of existence characterizing 'Il numero 1' came to the fore. The concluding section highlights three areas of improvement for the experimentation of creative-activist mode of existence with a vulnerable group.

Activism, translations and the ethnographic novel

The active role of research

The questions underlying the spatio-temporal politics outlined in the introduction are of a research-activist kind. This is not the space to provide a review of the extensive literature that our discipline has produced in the last two decades on this matter (The Autonomous Geographies Collective, 2010; Blomley, 1994; Mason, 2013; Routledge, 1996). It is enough to say, following a recent contribution by Derickson and Routledge, that research-activism can be thought of as a political ethos, guiding ideas and practices that are concerned with an 'attempt to find, generate, and resource potential rather than only provide intellectual critique' as well as 'to contribute to practices that are aimed at social transformation rather than merely the production of knowledge or the solving of local problems' (Derickson & Routledge, 2015, p. 6). There are at least three concerns that 'Il numero 1' shares with the research-activist ethos.

The first resides in asking: What do research constituencies gain from our research, and how should we make our findings relevant to them? According to Flyvbjerg, this question should define the role and meaning of social sciences: 'The purpose of social science is not to develop theory, but to contribute to society's practical rationality in elucidating where we are, where we want to go, and what is desirable according to diverse sets of values and interests' (Flyvbjerg, 2001, p. 167; for a review, Lancione, 2013b). Researcher-activists have argued similar points in the case of vulnerable groups (Pickerill, 2014) and around practitioners working with them (Mason, Brown, & Pickerill, 2013). At the most basic level, 'giving something back' to our research constituencies is the easiest response to these concerns, and it is arguably the one that guided the first phases of 'Il numero 1'. But who decides what that 'something' is? Is it still a prerogative of the researcher or a more mutual – and possibly empowering – endeavour (Pain, 2003)?

The second concern is about how to meaningfully engage with our research constituencies. Participatory Action Research (PAR) (Mason, 2015), Solidarity Action Research (SAR) (The Autonomous Geographies Collective, 2010), and also certain expressions of auto-ethnography from 'below' (Reed-Danahay, 1997) are all approaches working in this sense, filled both with limits and opportunities (Pain & Francis, 2003). The concern here is about positioning, a point at this project's core. The kind of positioning advocated by research-activism – and shared in the making of 'Il numero 1' – is not only about being reflexive on the position of the research into the field (McDowell, 1992), but about constructing horizontal solidarities and the blurring of hierarchical boundaries between researcher, researched and research (Rose, 1997). The point is not to dismiss the power imbalances but to acknowledge these hierarchical structures and then challenge them through participation, encounter, careful listening and co-production (Butz & Besio, 2009). Here, 'Il numero 1' is analysed, both as a practice through which some of these spaces were produced, and also as an artefact potentially able to produce further iteration of these spaces.

Third, as it is the case with any kind of activist endeavour (Chatterton & Pickerill, 2010), research-activism is a messy and emotionally demanding activity (Brown & Pickerill, 2009). The concern arising from this point is related to one's own well-being as well as that of others. Brown and Pickerill highlight in this regard the distinction between the space of the self – 'where we make sense of why we feel certain ways, and where we need to understand internalized oppression, in order to resist aligning with existing oppressive behaviours in society' (2009, p. 31) – and the interpersonal space, where 'through activist practices (such as ritual, material vulnerability, and laughter) and experiences of solidarity, the feeling of self grows, and emotional energy is shared' (2009, p. 32). Finding the balance between these two spaces is hard and demands a particular ethical *ouvre*, one aimed at cultivating the interpersonal milieu through which the activist struggle can be emotionally sustained. To me, ethnographic writing and, in particular, the ethnographic novel served as a way of enacting that milieu, by means of sharing and presenting, discussing and imagining, crafting and performing. The book became in this sense not only a research-activist exercise but also a ground for a shared and empowering kind of emotional labour (for a similar conclusion on storytelling, see Christensen, 2012).

These concerns raised by the research-activist literature have deeply affected my reasoning on how to be an engaged academic today, but there is something still largely absent from this literature: namely, an account of the role of non-human agencies in shaping activist endeavour (for a partial exception, see Askins & Mason, 2012; Askins & Pain, 2011). At the risk of oversimplifying, the narrative emerging from many of the works cited above is one that sees the researcher and their collaborators as carriers of all actions: They cooperate, co-produce, share, propose, present, accept, engage and more. But what if our activist endeavours are mediated in ways that are currently unaccounted? What about those agencies that are not human? Do they only allow 'us' to cooperate, co-produce and share, or do they portend more? If non-humans have a role in shaping the course of action (Latour, 1996) and the outcome of our political intent and practice (Amin & Thrift, 2013), why don't we take them more seriously, perhaps taking them into the calculus of our political strategies from the outset?

Translating the field

There are at least two benefits of bringing non-humans fully into the picture of research-activist scholarship. The first is still largely rooted in the classical Science and Technology Study (STS)'s understanding of non-human agencies (Callon, 1986; Law & Callon, 1992). To put it briefly, in recognizing those agencies, one will be able to assess not only when and where particular practices and engagements succeed or fail, but also to trace in detail how they do so. The second major benefit of considering non-humans in research-activist scholarship is that in doing so, new ethical demands and political questions could become apparent. For instance, by taking into consideration the small devices, discourses and myriad of agentic powers that populate the 'homeless city', one is able to challenge canonical narratives of personal culpability directed at homeless people. As the performative scholarship on homelessness shows, this is because the responsibility for one's own actions can then be seen as part of a wider network of agentic powers, not as the simple expression of one's own will (Cloke, May, & Johnsen, 2010; Lancione, 2013a).

The politics here are of 'cosmos' (Latour, 2004; Stengers, 2010), namely, of connections among multiple entities that can be understood only in their relations and are politically relevant only in their relations (Lancione & McFarlane, 2016). Thinking cosmopolitically about activism is then about enlarging the 'cosmos' that one currently takes into consideration to include the agencies and 'modes of existence' that canonical social theory has neglected for too long. The key is to trace how these modes of existence are assembled and to register what they can bring to the fore, both in terms of their actual capacities and potential (Anderson, Kearnes, McFarlane, & Swanton, 2012). One tool that may be implied in this sense is the notion of *translation*.

According to Actor-Network Theory (ANT), translation stands as a signifier of those human and non-human affiliations that move in the direction of – and are moved because they are fuelled by – a shared 'interest' (Callon, 1986; Murdoch, 2006, p. 76). Translation is not a 'thing', nor a 'power', but 'processes of negotiation, representation and displacement, which establish relations between actors, entities and places' (Murdoch, 1998; p. 362). Latour is clear: A translation is 'neither one actor among many nor a force behind all the actors transported through some of them *but a connection that transports,* so to speak, *transformations*' (Latour, 2005, p. 108) (my emphasis). It is worth stressing that the 'shared interest' of a translation should not be read from a human-centred perspective. It is not about what 'I' wish. Rather, the shared interest is constituted by the alignment of the network or, in other words, by the set of concatenations by which certain things are made to function in a certain way.

The reasons why I refer to the ANT's declination of translation can be summarized as follows. First, translations are always a matter of human and non-human affiliations. As the analysed case will show, the design of an artistic artefact, the materiality of collaboration, the activation of spaces and propositions all matter in defining the meaning and political capacity of the process itself. Second, the shared interest of translation must not be taken for granted, since that is only how the network aligns temporarily. In this sense, translations must be taken care of *actively*: They require a commitment that, to me, resonates with the research-activist ethos outlined above. Third, invoking translation serves the purpose of avoiding a simple celebration of human capacities in the fulfilment of the 'interest'. If one is serious about a cosmopolitical take on activism, then the story of 'Il numero 1' cannot be narrated only through 'Michele' and his abilities in narrative writing. Rather, it is about a wider agentic network that includes 'Michele', homeless people and the mechanosphere of Turin and thousands of small mediators co-producing that knowledge (McFarlane, 2011). In the end, talking of translations allows research-activists to reconsider the role of non-humans in their endeavours, to be committed not only to the 'cause' but to the wider agentic power of the network, and to reflexively de-centre their role in the mode-of-existence they are temporarily part of.

The ethnographic novel

At the core of the specific form of research-activist mode-of-existence of 'Il numero 1' there is an ethnographic novel, a form of fictive writing that could be defined, paraphrasing Geertz (1980), as one of the many 'blurred genres' characterizing contemporary social science: a mix of literature, poetic sensitivity and thick description through which one re-articulates how she/he writes about research. In short, the ethnographic novel is storytelling constructed around and incorporating both ethnographic evidence and elements of fiction (Crapanzano, 2012; Laterza, 2007).

Ethnographers have always published 'literary works alongside their ethnographic works' (Langnes & Frank, 1978, p. 124). From the first alleged ethnographic novel ever written in 1890 – 'The Delight Makers', Bandelier's story of the Keresan Pueblos, to the late Duke University Press' book by Tobias Hecht, 'After Life: An Ethnographic Novel' – the history of the discipline is punctuated by works of this kind (Schmidt, 1984). Although full recognition of this genre is yet to come, today, ethnographic fiction is more accepted as a legitimate way of communicating research outcomes. In a recent contribution, Jacobson and Larsen correctly point out that the post-structuralist turn in anthropology, which shifted the discipline away from more traditional forms of 'ethnographic realism', has contributed in this sense (Jacobson & Larsen, 2014). *Cultural Anthropology* has gone as far as saying that '[t]hough we've separated the fiction from the anthropology, there is no way to easily demarcate where fiction ends and anthropology begins' (Byler & Iverson, 2012). Indeed, if traditionally the distinction between the two had been based on the objective nature of the ethnographic 'truth', today that kind objectivity is largely gone. What is truth? As the same authors spell out: 'Aren't things like love, grief, shame, embarrassment and joy "true"? Does the "truth" even matter if it is ultimately useless to the communities that are studied and represented?' (Byler & Iverson, 2012).

Having said this, it is important to maintain a distinction between pure ethnography (which is a specific kind of writing) and ethnographic fiction: the latter would not be possible without the former. More importantly, it is pivotal to separate pure fiction from the ethnographic novel. An ethnographic novel is written on the basis of ethnographic research, with the aim to disseminate the findings of that research to a wider audience: It is explicit about where it comes from, and it has a clear sense of direction. Narayan has proposed four points to help one discern between ethnography and fiction: '(1) disclosure of process, (2) generalization, (3) the uses of subjectivity, and (4) accountability' (Narayan, 1999, p. 135). In the following sections, I will discuss my own ethnographic and collaborative book to argue in favour of a responsible approach to ethnographic fiction, one that takes Narayan's points seriously and is explicit about its spatio-temporal politics of representation.

Translations, or, the unfolding of a mode of existence

I undertook fieldwork with homeless people in Turin in 2009 and 2010 (with incursions in 2011 as well). The fieldwork consisted of spending time with (mostly) Italian homeless men while following their street journeys across the city, in volunteering at two Faith-Based organizations by delivering services to them and in undertaking observation at a drop-in centre managed by the municipality. The aim was to trace how the experience of homelessness was constructed in daily entanglements with city's spaces, services and artefacts (Lancione, 2013a). I started to write fictive accounts of my fieldwork's notes both to make sense of them and to be able to discuss them with some of my homeless friends. As I said in the introduction, these notes emerged in response to their demand for immediate 'outcomes' and 'impact'. From the accumulation of those notes, the idea of the book gradually emerged in mid-2010 and, after lots of work and entanglements, the book was published by Eris Edizioni in June 2011.

The mode-of-existence characterizing this endeavour, which I aim to evoke in the remaining parts of the paper, is the enmeshment of two agentic forces: 'my own', as a human being – with my capacities and wishes – and the wider mechanosphere of human and non-humans that the network gradually started to enrol. In the process, my agency was gradually marginalised in order to give room to other actants in the production, assemblage and extended life of the book. The final product is an artefact that, because of the way it has been written and the way it has been presented, carries with it the potential of producing further translations.

First translation: from the city to the structure of the book

A relevant aspect of the first translation that 'gave birth' to 'Il numero 1' was the participatory ethos foregrounding it. Since the very beginning, I shared my narrative notes with the group of homeless friends with whom I used to spend more time. These included Antonio, Amos, Pancrazio, Paolo and a number of other individuals around which the in-depth ethnography of my PhD thesis was based. Their first reactions to my writing, which were mainly of praise and excitement, filled me with a renewed spirit: I felt, for the first time after months of fieldwork, that I was moving towards a possible answer to Antonio's original challenges. Another response that the feedback generated was a sense of increased expectation in my regard: my homeless friends now knew I was working on something and, with little provocations here and there, they were always ready to remind me of my 'task' (to 'tell their stories').

The affective charge that this emergent mode-of-existence was bringing to the fore exceeded our human-based interaction. In other words, that interaction took place in different spaces characterized by heterogeneous affective ambiances that mediated the way we ended up doing the book (Anderson, 2009). Sharing notes and chatting with homeless people about the book in the institutionalized setting of the Vincenziani's soup kitchens, while they were queing for the free distribution of clothes, or waiting to receive counselling, definitely played a role in facilitating the expression of their resentment for certain aspects of the service. The buzzy atmosphere of a small café close to the train station, where I used to chat with Amos; his phone calls always interrupted by all sorts of emergencies and his numerous drafted emails containing chunks of his introductory essay, played a role in defining the patchworked nature of our collaboration. Eleonora's (the illustrator) messy desk and her numerous, thick sheets of white paper, criss-crossed by black strokes; my tiny desk in a studio in Turin, where I sat and wrote the whole thing almost in a row; the living room of one of the book's publishers, filled with manuscripts, cigarette butts, red strokes on white paper and a huge cat looking at us at all times... All these things, with their immanent power and affective charge, moulded the way we wrote, assembled, designed and eventually published 'Il numero 1'.

I am not recalling these spaces and atmospheres here to romanticize the process of doing the book but to ground it and make it real. It was indeed through those spaces, because of their specific settings and affective ambiances, that our network and mode-of-existence started to take form and substance. In this regard the city itself, Turin, played a fundamental role in shaping the overall structure of 'Il numero 1'. It is because of the complex entanglement of public institutions (the Municipality of Turin), private ones (mostly Faith-Based Organisation), vulnerable people (such as the homeless), the general public, moving on the plane of a post-industrial and stratified city that a bare novel wouldn't have been enough to achieve the politics of representation (interest) grounding our mode-of-existence. In stating this, I am not dismissing the rational choice of proceeding with a composite volume; but I am stressing the *mediated* nature of that rationality. That mediation is the process by which Turin entered in the book – through it stories, institutions and entanglements – while we, at the same time, were trying to connect the book back into the city – structuring it in such a way for the plug to be possible. The preface by Amos and the body of the 'ethnographic novel', Eleonora's 20 black and white drawings and the 'appendix' are not, in this sense, a purely rational choice but the elongated spaces of that city and its atmospheres. The city called, demanded and largely imposed to us the composite book that we ended up doing. It comes without saying that we could have ignored that call, but this is precisely the point of thinking through translations: to allow for complexity to room large and not been ignored, controlled or reduced.

Second translation(s): prefigurative devices

If the first translation took place in the mingling of bodies and the city, giving rise to the network and the structure of the book, the artefact of the book came to the fore with a second set of translations. These could be defined as 'prefigurative', since they comprise a number of arrangements that were put in place in order to connect the book to its readers in meaningful ways (namely able to activate that space of encounter and time of immediacy orienting the network's interest). Once again, I understand these arrangements as mediated processes, or post-human devices, which encompass at least three areas: how to write, how to illustrate and how to design the book.

The device of narrative

Starting from the book cover, the aim of the novel is to represent the post-human process of becoming a homeless subject (Figure 1). The plot of the novel is therefore centred on a character who, having lost his means of sustenance (family, home, job), gradually becomes home-*less*: the process by which he becomes entangled with the time, spaces and subjects populating the 'homeless city', which include other homeless people but also particular Turin contexts, such as shelters, soup kitchens, the black market, and the train stations. Each of these entanglements unfolds on the basis of real-life events drawn from my participant observation data. Laid out in succession, these events show the process experienced by the vast majority of my informants: a shift from one form of normality (householder, worker, parent) to another (nomad, hustler, loner).

The experiences collected through my ethnography demanded, in this sense, specific stylistic choices to be made (the ethnography acted, in this sense, as mediator of those choices). The major aspect in this regard concerned the decision to leave the main character nameless. He speaks and thinks, but his name is never voiced, and nobody ever asks for it. The purpose of this device was to give a sense of 'collectivity' in the narrated episodes. These are indeed carved out of many life-experiences but are not specific to anyone in particular: they are part of the normality of homeless men's lives in contemporary Turin. Another important choice was to reduce dialogues to a minimum and to give ample resonance to inner thoughts and feelings, in order to portray what people told me during many hours of walks and chats, as well as the atmospheres I myself confronted during those walks and chats. In letting these voices come together in the thoughts, and even dreams, of the three main characters, I over-simplified them. However, I believe that these devices allow the book to bring the reader 'closer' to the collective experience of becoming a homeless-subject, and thus better evoke the feelings, fears and hopes of these people (Robinson, 2011). Last, since small objects and places had an enormous importance in my understanding of homelessness and in crafting the experiences of the homeless people I encountered, 'Il numero 1' contains at-length descriptions of apparently insignificant details – like sidewalks, parks, streets, clothes, urban landscapes and so on. Taken together, these stylistic devices aim to render the experience of homelessness in Turin emotionally intelligible to the reader.

The following excerpts are two examples of these devices. In the first, the hero confronts the issue of not having a formal residence:

> I was alone, it was cold and I had no residence. That's the thing. With the residence I could have stayed a month in the same public dormitory, and I wouldn't have had to wander around with my things from place to place, day after day, hour after hour. I ended up hating the Identity Card too. I was looking at it when I was queuing at the soup kitchen. When I was cueing at the dormitory. When I was queuing at the employment agency. I'd take it out of my paper-stuffed wallet – numbers, holy pictures, and little money – and I'd stare at it. I saw myself in it, but distorted. That face didn't belong to me any more. It was part of a story that I was slowly forgetting. And not only that: it stated that my residence wasn't in that city. My residence wasn't in any fucking place. I wandered. But if you want to sleep comfortably, you can't do that. [...] And then, I'd

Figure 1. The book cover. Source: Illustration by Eleonora Mignoli.

tuck my ID away, I'd light a cigarette and I'd look around feeling myself, very slowly and gently, slipping away. (pp. 41, 42)[1]

This passage offers a glimpse of a major issue that homeless people faced back then in Turin, namely the fact that in order to access the public shelters for more than a night, one had to have a legitimate proof of identity released by the City. More importantly, using the hero's train of thought as the main stylistic device, one is able both to portray the power of that municipal rule in shaping subjects' feelings and emotions, and the role that a small object (an ID card) can have in triggering powerful affective responses. The following passage offers a similar example related to the distribution of clothes by Faith-Based Organisations (FBOs) in the city. Once again, through this particular writing style – which mixes thoughts, emotions and the materiality of life at the margins – I try to connect the intimacy of clothing, both to the limits of a particular service provision and the struggles of becoming-homeless. What (hopefully) the reader gains from this is a sense of how the experience of homelessness is constructed in the micropolitics of everyday life:

> I would go to this soup kitchen, eat and then wait for my turn to get the clothes. A couple of shirts. Shoes. Mostly underwear: I had to throw it away once a week because in the dormitories there is no washing machine. One can't imagine how pleasant it is to wear clean underwear till you have longed for it as much as a bed to sleep in and a roof over your head. It's one of the few items of clothing that they gave us new, unlike all the others that were second hand. However, quite often they would give us the underwear without wrap, without package. The reason was that they didn't want us to go and sell it. It's understandable, I thought. But still, it was an insult, a lack of respect. It shattered the last fort that still made us feel men, worthy of briefs. Give me the wrap, let me touch it. I want to feel the plastic under my fingers. I want my hands, not yours, to be the ones that break the soft shell of the new pair of underwear, folded and clean. I want to bury my nose in the cotton, just for a heartbeat, before slipping it on. What divides the man from the beast is not the faculty of reason or free will. It's underwear. The fine line it's all in the briefs, in owning a clean pair. And the more dirty they get – farting, with the help of bad milk, does the trick – the more the line becomes thinner. It becomes a solid barrier, very solid. The next step is starting to bleat, clapping your hands on command and rummaging with dirty nails through a garbage bin. I knew, because I saw, that some people did exactly that. And I was certain, when I spotted them, that their underwear would be nothing but a tattered piece of fabric smeared with the dirt of their own shit. I couldn't bear that sight. I would not resign myself to that destiny. I wanted to kill them, all of them. And to throw, plugging my nose, their underwear to the dogs. At that time my underwear was still sweet, white and soft. Nice. Clean. (p. 47)

The prefigurative role of narrative writing is also clear in the preface written by Amos. There Amos offers a brief but poignant account of his experience as a homeless man in the city, which resonates with that of the main character of the novel, although using a more immediate, heartfelt tone. For instance, talking about the disconnection between one's own desires and the normative approach adopted by the City's services, he writes:

> The most unsettling thing about this matter is that I've never heard a social worker say C'mon, you're gonna find a job! Or try to go there... or give any other advice of this kind; everyone's single worry was that you'd get in in time for the service, and they gave you guidelines only on how to sign up for a new centre or dormitory ... (p. 10)

The tropes evoked by Amos are both material ('sign up'; 'guidelines') and emotional (the refusal of the social workers), and they speak loudly of the inadequatedness of the relationship between service-providers and homeless people, of the lack of meaningful interaction and care. In other words, they speak truth to power in a way that most people having an understanding of homelessness can get. Playing on similar chords, the final essay of the book was addressed to a sub-set of readers: policy-makers and practitioners working in the homeless domain. For this reason, the essay is more scholarly, following a clear analytical path that identifies specific issues within Turin's institutions and policies. The role of this closing chapter is to work as device able to connect the content of the novel to the politics of homelessness at stake and to advance concrete proposition for actions.

The device of illustration

The choice of inserting illustrations in the book was mediated by three concomitant agencies. The first was that of the numerous paintings, drawings and books about illustrations that populated the flat that Eleonora and I were sharing in Turin in late 2010 and 2011, which were acting as a constant reminder of the power of images in conveying meaning and opening lines-of-flight (Berger, 1972). Second, there was the wish of Eleonora to learn more about the story narrated in the book and our idea – as a couple – to share the affective ground of a common project. Third, there was the explicit opportunity, given by the people from Eris Edizioni, who worked with us on the manuscript, to leave room in the book for a number of evocative drawings: pages were added to the final count, more money was allocated to the project and a new channel of communication was made possible.

Eleonora's illustrations are dense, filled with machinic details, portraying the entanglements of matters, bodies and hopes/fears/dreams that make up the 'homeless subject' narrated in the novel. Take, for instance, the illustration in Figure 2. Several homeless people have told me that it perfectly captures the realities of sleeping in abandoned train wagons during the winter: the cold, the solitude, the harshness, the fear, and the dreams of warm beds that turn into nightmares.

Eleonora's artwork emerges from her personal engagement with some homeless individuals included in my research but also from her close reading of the ethnographic novel. As she puts it:

> It's a peculiar writing, very dry yet precise, full of short sentences, as if told by someone out of breath. In the tight gap between paragraphs I felt that my illustrations could find their place. My desire was to create a sort of counterpoint: where the novel takes you urgently from one word to the other, the illustration creates a saturated space where the mind can rest and wander. At the same time, with the sharp B&W, I wanted to keep the mood of the novel.

A clear example of this visual style is represented by the first illustration of the book, which intersects with the second chapter of the novel, where the hero dwells on memories of his recent past but has yet to be revealed as homeless to the reader. This is how Eleonora interpreted this passage, which leads her to draw the table in Figure 3:

> The first illustration shows a balloon hosting on its surface a small microcosm. A hand holding a needle approaches, ready to pop it. It's the beginning of the story, where the protagonist falls out of grace. He lives at the top of the balloon, in the house with the swimming pool on the background of a growing environment, doesn't see the needle approaching nor knows that between them and nothingness there's only a thin layer. But we know that when the balloon pops they will be violently ejected out of that existence and everything will fall around them.

These and other drawings in the book worked as devices able to connect the actuality of things narrated in the story to the potentiality of escape routes, dreams and affectivities that the written word is not able to express in full. Illustrations are, in this sense, key to perform the second translation of the book: in making it more intelligible, alive and real (in a Deleuzian sense, both actual and in potential; Deleuze, 2001).

Figure 2. Sleeping in the train station. Source: Illustration by Eleonora Mignoli.

The device of design

The design of a book is a quintessential matter for a marketing strategist. But it is also very important from an activist-research point of view, when the aim is to allow the artefact to

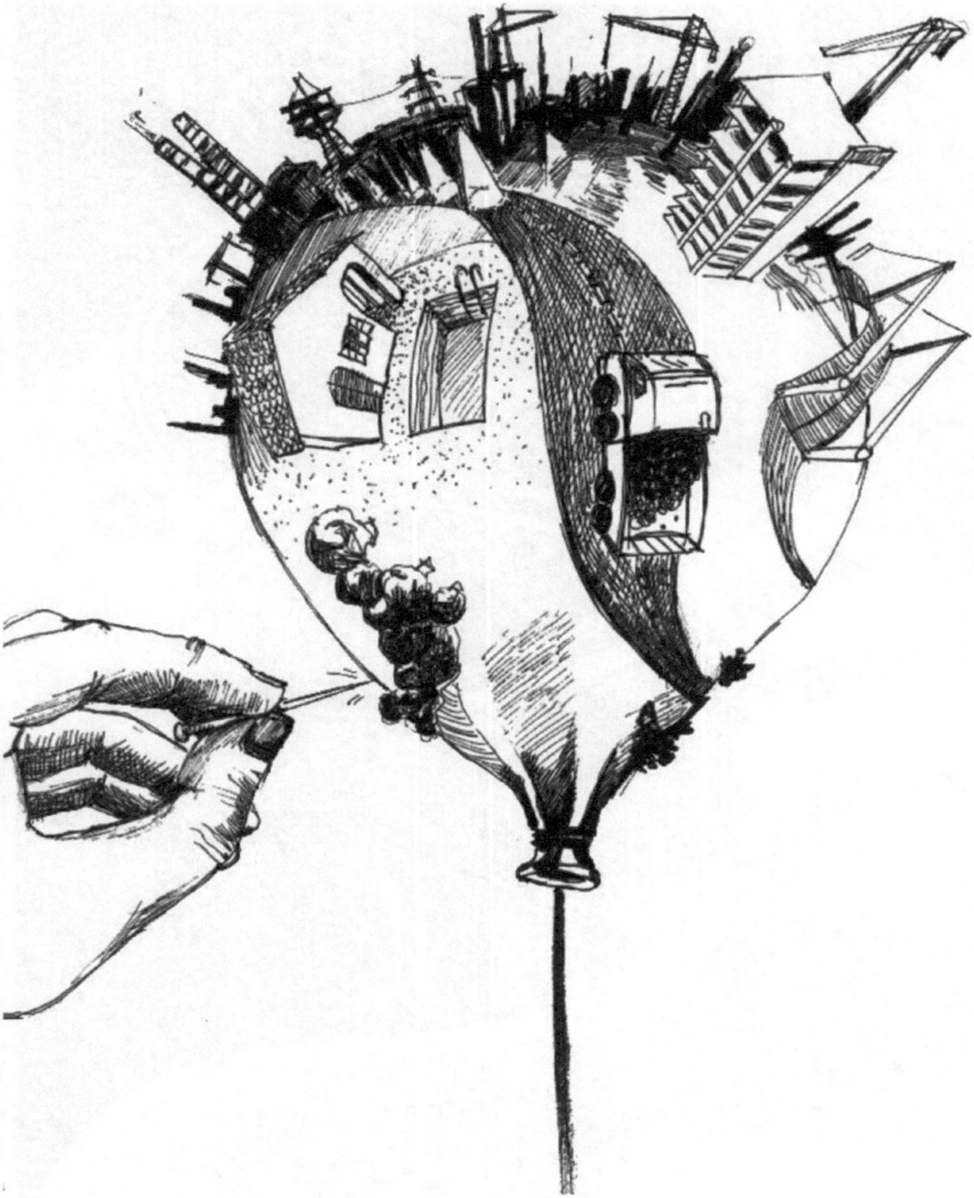

Figure 3. The beginning of the story. Source: Illustration by Eleonora Mignoli.

be able to sustain itself, to meaningfully engage readers, to travel and to perform future translations. In this regard, there are a number of points that I want to briefly highlight concerning 'Il numero 1'. The first regards the size of the book. Since we wanted something easy to bring along, the Publisher proposed a rather small paperback format that could sit in a jacket's pocket. Moreover, this choice was also aligned with the idea of producing a very affordable book (10 euro), printed on recycled paper but, at the same time, of a good quality in its finish. The latter was particularly important, since the appearance of a book already contains in itself the power to communicate trust, to solicit interest and to provoke curiosity.

By mid-2010, I already had a full draft of the manuscript in my hands. Although the draft was still very provisional at the time, I considered distributing it among my homeless friends as it was: a pile of self-printed stapled sheets. However, I realized that the pile of sheets was lacking a material form and structure that would allow it to speak more prominently and effectively to its potential audience. This is why, in the end, I submitted it to three different publishers. Although I received offers by all of them, I finally accepted only the one by Eris: the only radical publisher among the group, and the only one from Turin.[2] More than one year passed from the moment I finished the manuscript to when it was published but the outcome was a well-designed artefact, able to stand on its own.

The third and last point concerning the role of design in promoting future translations of the book lies in the choice of its publisher to license it under a Creative Common License. While protecting its authors, this form of licensing is designed to allow anyone to make copies of the book for free, without incurring any form of liability. By enrolling the book into the Creative Common network, we provided it with further capacities to make an impact, especially when it comes to its diffusion among vulnerable and non-financially able groups, such as homeless people.

Third translation: engaging in a mode of existence

The second set of translations just outlined was of a prefigurative nature. Through the agentic powers expressed by its structure, content, style, illustrative material, craft, open-source policy and price, the book was made freer, more able to perform future translations and, hence, more apt to pursue its short and long-term politics. Arguably, however, writing and publishing the book was only a starting point for the achievement of these ends. A further layer of engagement needed to be brought to the fore, one able to fully 'activate' the activist mode of existence of 'Il numero 1'. Aside from interviews with radio and local newspapers, and the recognition the book received within literary circles,[3] there are three ways through which the actor-network of the book pursued this task.

The first related to the promotion of the book among the homeless community. After an initial distribution of copies to the people with whom I was close, Amos took responsibility to spread copies of the book among his peers. If I had already shared excerpts from the text with some members of this network, for others the book came as a surprise. Between the days in which the distribution took place and the moments when the first comments started to flow, I felt terribly stressed. The fact that the book was out there re-positioned me in unpredictable ways in the eyes of the people with whom I had spent so much time. The agentic power of the book was totally dissociated from 'Michele': I could not control it, although I had my name on its cover. After a while, feedback started to flow. They were, for the most part, enthusiastic. For example, Antonio told me: 'I like it, I like it a lot. Good on you! Cardù [a character in the book loosely based on him] is indeed smart, he is almost as smart as me. But you know that I am smarter!' Other comments read:

R., reported by Amos: R. told me that he read it four times. He knows all the passages by heart! He also said that although you tried to make the ending less sad, he likes it anyway.

A.: I like the book! It seems my life … But do you really think someone is going to read it?

P.: 'You did it! Bravo! I am going to tell everybody about it because here there are too many things that do not work. The font is too small though... I need to put on my glasses!

P.: Great Michi! The book is cool! I didn't think you could write it like this. I did not like the sex scene though. But all the rest is great, it is true. I want to write mine now!

Homeless people who read it generally liked the way the book portrays their lives, and they told me that they could see themselves in this or that aspect of the overall narrative. At least one comment, however, wasn't totally positive. Angelica, a homeless woman at the time of my study, told me:

> I like it, I mean, I can see some things of my life in there ... But I don't like how it is all so sad in the book. You did not say about the fact that life is beautiful, that there is hope! I am going to write these things in my book now!

Although the book terminates with a hint of hope, Angelica was right in pointing out the overall pessimist tone of the writing. Both the book and the fieldwork fail to take her experience, and surely that of many others, fully into account.[4]

This initial engagement with homeless people set the stage for a series of eight public events in Turin co-organized by the editors, Eleonora, Amos, a number of other people and myself. These events included presentations in bookshops, bars and social clubs that took place in July 2011. Each one of these was structured around my reading out some book passages and some newer texts, Fabrizio, a musician friend, accompanying my readings with selected songs that evoked the themes of the book and Eleonora painting live throughout the performance then dividing the paintings into small pieces that were distributed to the public (as further material afterlife of the tales).[5] Friends, homeless people, colleagues from the University of Turin and a wide number of people simply interested in live performances attended the events. Although we did not collect systematically people's responses, their comments were generally very positive. I believe the mix of music, painting, words, alcohol, bodies, chairs, books, colours, hopes and ideas allowed the book to speak to a broader public in effective ways. Indeed, the homeless people who were able to attend the events said they found them powerful and were genuinely gratified by the experience.

The third level of engagement took place with practitioners and policy-makers. Since the appendix of the book was targeted at them, the first action was to put it online as a free downloadable document. I then forwarded it to all my contacts, also inviting them to the public presentations, and I mailed copies of the book to them as well. Through the possibilities offered by the web and the sharing and forwarding of original emails and posts, the network of the book expanded its reach. Many young social workers and practitioners working in the homelessness sector in Turin came to the events, showing appreciation and support. 'Scarp de tenis' – the main journal treating homelessness-related issues in Italy, edited by a range of organizations working in the field and written and sold by homeless people themselves – published a piece on the book, exposing its message to a wider group of practitioners (the same happened with the national newspaper *La Repubblica*, which reported on the book).

Although the book itself could not change the politics of homelessness in Turin, it did contribute to a debate around the quality of service provision that was already in place at the time. An example in this sense is the Vincenziani's soup kitchen where I volunteered and undertook observation during the fieldwork. Two nuns in particular were very open to my critical observations and involved me in consultations leading to the partial re-design of their service provision. One of them also came to a public event, criticizing me for the use of dirty language in the book, but appreciating it overall. On the municipal side, 'Il numero 1' reached some practitioners, thanks to the relationship that I maintained over time with a few of them. Despite his late engagement with the book, M.D. (the director of city services for adults in need) saw it as a tool to set off a discussion with his colleagues. As he wrote to me:

> I finally read […] 'Il numero 1'. It is truly impressive how the book manages to capture the reality of the services offered by the City of Turin, and of homeless people, at least according to my experience. I hope to be able to use the Appendix in the reflections I will propose to my colleagues, both in City and in the cooperatives'. (M.D., e-mail communication, May 2012)

Conclusions: on ways of improving translations

With this paper, I argued for the need to write more, better and differently to engage with radical praxis today (Mitchell, 2006), via an analysis of a concrete case in which, at least to a certain extent, a non-academic text has allowed for a meaningful re-appropriation of academic research. In this sense, the paper has shown how researchers using ethnographic methods can utilize creative strategies to pursue goals that sit at the crossroads of research and activism. I also argued that research-activist scholarship should take non-human agencies seriously, since the latter play a foundational role in the way we conceptualize, perceive and perform activist endeavours. To show this, I analysed a collaborative book project called 'Il numero 1', understanding it as a mode of existence, namely as a modality of being and acting in the world. That modality consisted of an extensive network of humans and non-humans sharing the same 'interest' – which I conceived, in ANT terms, as the peculiar orientation of that network. In the case of 'Il numero 1', the network was oriented towards the fulfilment of specific spatio-temporal politics of representation, aimed in the short term to empower homeless people – allowing them to become part of the politics of their representation - and in the long run to tell their story through an opportune, powerful, medium.

As a way of concluding this paper, I want to reflect on three shortcomings of the project. These are taken from the current literature on research-activism and cover also the less satisfying aspects of 'Il numero 1'. First, in the book, there is a problem of *commitment* with policy-makers. 'Il numero 1' has been a good exercise for listening and reporting the stories of homeless people and their pleas. However, it entirely failed at listening and reporting the stories and the pleas of service providers. The appendix of the book, although addressed to them, does not take their perspectives into consideration: it tells them what to do and how to do it better, without fully discussing the limits and opportunities of their operations. Paraphrasing Derickson and Routledge, the research underlying the book could have been more resourceful if it had been 'designed explicitly to ask and answer questions that non-academic collaborators want to know' (2015, p. 1).

Second, the book suffered from a lack of full *co-production*. Although the book was factually assembled by at least four people – Amos, Eleonora, the people from Eris and myself – plus a number of other actants and actors, and even if notes were shared with homeless people and feedbacks taken seriously into consideration, a more progressive co-production is needed to achieve even more empowering outcomes (Lassiter, 2001).

Lawless has spoken in this sense of 'reciprocal ethnography', the process through which written text is continuously exchanged and commented upon between the researcher and its collaborator, in order to arrive at a form of deep conjunct writing. Such a process requires an interrogation of the hermeneutic circle of doing fieldwork and writing ethnography. As Lawless puts it:

> 'I have not relinquished my role as interpreter, as thinker, as objective observer. But I have given up the notion of scholar voice as privileged voice, the scholar's position as more legitimate because it is the more educated or more credible one'.(Lawless, 1992, p. 312)

Last, and related to the previous point, 'Il numero 1' could have benefited from a more explicit participatory approach to its afterword. The problem here is one of *co-action*: although homeless people were invited to the live performances, not enough was done to actively engage them in its presentation and dissemination. The empowering force of the project was, therefore, somehow not fully exerted by a lack of a full co-ordination in production and action, points that need to come at the centre of future translations with vulnerable groups.

Recognizing these shortcomings, the paper shows how translating the fieldwork and working towards a research-activist mode of existence is a complex, tiresome and fragile endeavour. However, it is a necessary one that has the potential to make academic findings available to a broader public and may help scholars to 'give something back' to their participants in partial but situated and committed ways. In the case of a vulnerable group, this mode of existence can bring to the fore empowering experiences for research collaborators that may allow them to rethink taken-for-granted notions of what is possible and what is not. At the centre of this mode of existence, there is a cosmopolitical understanding of what it is to be an activist and researcher, one able to encompass standardized narratives of self-capabilities in order to focus on the affective, on passivities and on non-human agencies. In the case of 'Il numero 1', it is the attention paid at the agentic power of devices, such as narrative writing, drawings and designing, that made the book able to stand on its own and to endure translations that continue today.

Notes

1. All the text taken from 'Il numero 1' was translated from Italian by Eleonora Mignoli and revised by myself.
2. Eris Edizioni is an independent publishing house committed to support and produce alternative and radical knowledge, also through the adoption of Creative Commons licensing. It defines itself as being 'against any gender discrimination, anti-fascist, and against any regimes depriving any human kind from their rights' (http://www.erisedizioni.org).
3. The book has been positively reviewed by 'L'Indice dei libri del mese', arguably the most important literary magazine in Italy at the time. It also got a special recognition at the XXIV edition of the 'Italo Calvino' prize, a renowned literary competition for emerging authors.
4. From that criticism on, I then encouraged Angelica to write her own experience of homelessness, which eventually ended up published in a book called *The Other Face of Life* (L'altra faccia della vita).
5. Video of some of these performances can be seen at the following page: http://www.michelelancione.eu/narrativapoesia/video/il-numero-1-performance/

Acknowledgements

I want to dedicate this paper to Pancrazio and Paolo, two close homeless friends who sadly passed away in 2014 and 2015, respectively. To Antonio (not his real name), of whom I lost traces. To Amos, with whom we still share a lot and we still learn a lot from each other. And also to the city of Turin, which made all of this possible. Thanks to the people who supported the making of the novel, the Editors from ERIS, Fab-Puccettó and the friends in Turin, Roby for the support in post-fieldwork depression, Charlotte Kuhlbrandt, Eszter Krasznai Kovàcs and Alberto Vanolo for the chats around some of this paper's concerns, and above all, thanks to Eleonora 'Leo' Mignoli. The research beyond the novel and this paper wouldn't have been possible without the generosity of the Department of Geography in Durham, the DIST in Turin and the Urban Studies Foundation. Lastly, the paper has befitted from the very insightful comments of Nadia von Benzon and S&CG's referees and Editor, whom I thank very much.

Disclosure statement

This paper does not intend to publicise the author's novel, but to use that example as a way to critically assess the potential of creative methodologies in fostering alliances for social change at the urban margins.

References

Amin, A., & Thrift, N. (2013). *Arts of the political: New openings for the left*. London: Duke University Press.

Anderson, B. (2009). Affective atmospheres. *Emotion, Society and Space, 2*, 71–81.

Anderson, B., Kearnes, M., McFarlane, C., & Swanton, D. (2012). On assemblages and geography. *Dialogues in Human Geography, 2*, 171–189.

Askins, K., & Mason, K. (2012). Us and us: Agonism, non-violence and the relational spaces of civic activism. *ACME: An International E-Journal for Critical Geographies, 14*, 422–430.

Askins, K., & Pain, R. (2011). Contact zones: Participation, materiality, and the messiness of interaction. *Environment and Planning D: Society and Space, 29*, 803–821. doi:10.1068/d11109

Atooh, K. (2011). The bus hub. *ACME: An International E-Journal for Critical Geographies, 10*, 280–285.

Berger, J. (1972). *Ways of seeing*. London: Penguins.

Blomley, N. (1994). Activism and the academy. *Environment and Planning D: Society and Space, 12*, 383–385.

Brown, G., & Pickerill, J. (2009). Space for emotion in the spaces of activism. *Emotion, Space and Society, 2*, 24–35.

Butz, D., & Besio, K. (2009). Autoethnography. *Geography Compass, 3*, 1660–1674.

Byler, D., & Iverson, S. D. (2012). Literature, writing & anthropology. *Cultural Anthropology Online*, 1–6. Retrieved from http://www.culanth.org/curated_collections/5-literature-writing-anthropology

Callon, M. (1986). Some elements of a sociology of translation: Domestication of the scallops and the fishermen of St Brieuc Bay. In J. Law (Ed.), *Power, action and belief: A new sociology of knowledge* (pp. 196–223). London: Routledge & Kegan Paul.

Chatterton, P., & Pickerill, J. (2010). Everyday activism and transitions towards post-capitalist worlds. *Transactions of the Institute of British Geographers, 35*, 475–490.

Christensen, J. (2012). Telling stories: Exploring research storytelling as a meaningful approach to knowledge mobilization with indigenous research collaborators and diverse audiences in community-based participatory research. *The Canadian Geographer/Le Géographe canadien, 56*, 231–242.

Cloke, P., Cooke, P., Cursons, J., Milbourne, P., & Widdowfield, R. (2000). Ethics, place and environment, reflexivity and research: Encounters with homeless people. *Philosophy & Geography, 3*, 133–154.

Cloke, P., May, J., & Johnsen, S. (2010). *Swept up lives? Re-envisioning the homeless city*. Oxford: Blackwell.

Crapanzano, V. (2012). Interview with vincent crapanzano. *Cultural Anthropology Online*. http://doi.org/papers2://publication/uuid/219A2CD7-F298-4D5B-B8D9-80811FE43B14

Deleuze, G. (2001). *Pure immanence*. New York, NY: Zone books.

Derickson, K. D., & Routledge, P. (2015). Resourcing scholar-activism: Collaboration, transformation, and the production of knowledge. *The Professional Geographer, 67*(1), 1–7.

Eshun, G., & Madge, C. (2012). "Now let me share this with you": Exploring poetry as a method for postcolonial geography research. *Antipode, 44*, 1395–1428.

Flyvbjerg, B. (2001). *Making social science matter: Why social inquiry fails and how it can succeed again*. Cambridge: Cambridge University Press.

Geertz, C. (1980). Blurred genres: The refiguration of social thought. *The American Scholar, 49*, 165–179.

Governa, F., & Puttilli, M. (2016). After a revolution. Public spaces and urban practices in the core of Tunis. In M. Lancione (Ed.), *Re-thinking marginality: Assemblage, space and subjects* (pp. 42–59). London: Routledge.

Jacobson, M., & Larsen, S. C. (2014). Ethnographic fiction for writing and research in cultural geography. *Journal of Cultural Geography, 31*, 179–193.

Lancione, M. (2013a). Homeless people and the city of abstract machines: Assemblage thinking and the performative approach to homelessness. *Area, 45*, 358–364.

Lancione, M. (2013b). Truthful social science or : How we learned to stop worrying and love the bomb. *Journal of Political Power, 6*, 147–155.

Lancione, M. (2014). Assemblages of care and the analysis of public policies on homelessness in Turin, Italy. *City, 18*, 25–40.

Lancione, M., & McFarlane, C. (2016). Infrastructural becoming: Sanitation and the (un)making of life at the margins. In I. Farías & B. Anders (Eds.), *Urban Cosmopolitics* (pp. 45–62). London: Routledge.

Langnes, L. L., & Frank, G. (1978). Fact, fiction and the ethnographic novel. *Anthropology and Humanism Quarterly, 3*, 18–22.

Lassiter, L. E. (2001). From "reading over the shoulders of natives" to "reading alongside natives", literally: Toward a collaborative and reciprocal ethnography. *Journal of Anthropological Research, 57*, 137–149.

Laterza, V. (2007). The ethnographic novel: Another literary skeleton in the anthropological closet? *Suomen Antropologi: Journal of the Finnish Anthropological Society, 32*, 124–134.

Latour, B. (1996). On interobjectivity. *Mind, Culture, and Activity, 3*, 228–245.

Latour, B. (2004). Whose cosmos, which cosmopolitics? Comments on the peace terms of ulrich beck? *Common Knowledge, 10*, 450–462. doi: 10.1215/0961754X-10-3-450

Latour, B. (2005). *Reassembling the social*. Oxford: Oxford University Press.

Latour, B. (2013). *An inquiry into modes of existence. An anthropology of the moderns*. Cambridge, MA: Harvard University Press.

Law, J. (2008). On sociology and STS. *The Sociological Review, 5*, 65–649.

Law, J., & Callon, M. (1992). The life and death of an aircraft: A network analysis of technical change. In W. Bijer & J. Law (Eds.), *Shaping technology/building society: Studies in sociotechnical change* (pp. 21–52). Cambridge: The MIT Press.

Lawless, E. J. (1992). "I was afraid someone like you ... an outsider ... would misunderstand": Negotiating interpretive differences between ethnographers and subjects. *The Journal of American Folklore, 105*, 302–314.

Marston, S. A., & De Leeuw, S. (2013). Creativity and geography: Toward a politicized intervention. *Geographical Review, 103*, iii–xxvi.

Mason, K. (2013). Academics and social movements: Knowing our place, making our space. *ACME: An International E-Journal for Critical Geographies, 12*, 23–43.

Mason, K. (2015). Participatory action research: Coproduction, governance and care. *Geography Compass, 9*, 497–507.

Mason, K., Brown, G., & Pickerill, J. (2013). Epistemologies of participation, or, what do critical human geographers know that's of any use? *Antipode, 45*, 252–255.

McDowell, L. (1992). Multiple voices: Speaking from inside and outside "the project". *Antipode, 24*, 56–72.

McFarlane, C. (2011). *Learning the city: Knowledge and translocal assemblage*. Oxford: Wiley Blackwell.

Mezzadra, S., & Neilson, B. (2013). *Border as method, or, the multiplication of labor*. Durham, NC: Duke University Press.

Mitchell, K. (2006). Writing from left field. *Antipode, 38*, 205–212.

Murdoch, J. (1998). The spaces of actor-network theory. *Geoforum, 29*, 357–374.

Murdoch, J. (2006). *Post-structuralist geography*. London: Sage.

Narayan, K. (1999). Ethnography and fiction: Where is the border? *Anthropology and Humanism, 24*, 134–147.

Pain, R. (2003). Social geography: On action-orientated research. *Progress in Human Geography, 27*, 649–657.

Pain, R., & Francis, P. (2003). Reflections on participatory research. *Area, 35*, 46–54.

Pickerill, J. (2014). The timeliness of impact: Impacting who, when, and for whose gain? *Acme, 13*, 24–26.

Reed-Danahay, D. (1997). Introduction. In D. Reed-Danahay (Ed.), *Auto/ethnography: Rewriting the self and the social* (pp. 3–21). Oxford: Berg.

Robinson, C. (2011). *Beside one's self. Homelessness felt and lived*. New York, NY: Syracuse University Press.

Rose, G. (1997). Situating knowledges: Positionality, reflexivities and other tactics. *Progress in Human Geography, 21*, 305–320.

Routledge, P. (1996). The third space as critical engagement. *Antipode, 28*, 399–419.

Sandercock, L., & Attili, G. (2012). Unsettling a settler society: Film, phronesis and collaborative planning in small-town Canada. In B. Flyvbjerg, T. Landman, & S. Schram (Eds.), *Real social science: Applied phronesis* (pp. 137–165). New York, NY: Cambridge University Press.

Schmidt, N. J. (1984). Ethnographic fiction: Anthropology's hidden literary style. *Anthropology and Humanism Quarterly, 9*, 11–14.

Stengers, I. (2010). *Cosmopolitics I*. Minneapolis, MN: University of Minnesota Press.

The Autonomous Geographies Collective. (2010). Beyond scholar activism: Making strategic interventions inside and outside the neoliberal university. *Acme, 9*, 245–275.

Residential ethnography, mixed loyalties, and religious power: ethical dilemmas in faith-based addiction treatment

Andrew P. J. W. Williams

ABSTRACT
The paper provides a platform for geographical reflection on the hidden struggles ethnographers face working in the area of religion, addiction and drug treatment. Specifically, it examines the complex ethical and practical dilemmas involved in residential ethnography inside a faith-based therapeutic community working in the area of addiction and rehabilitation. Residential ethnography provided valuable insights into social life in therapeutic community, and more broadly, offers an ethical and participatory approach to research in closed institutional settings. Residential immersion in faith-based therapeutic environments however raised significant challenges around identity management; access and consent; and the dilemma of 'mixed loyalties' – a term that describes a set of ethical practices characterised by ethical conflict, compromise and negotiation in which the researcher, by nature of their participation, is expected to conform to certain values, practices, and procedures that may contradict their own personal ethics. To ground discussion on the variegated and contested nature of mixed loyalties, this paper examines the exercise of religious power and the ways ethnographers become enrolled in, and must negotiate, a series of power-dynamics that are unclear, uncomfortable, and potentially exclusionary. By illustrating the difficult decisions ethnographers must make when negotiating pressures to uphold – or challenge – religious beliefs and practices in faith-based addiction treatment settings, this paper calls for greater critical reflection on the ways geographers are implicated in the field and the practical ethics of engagement used to navigate ethical tensions.

Ethnographie résidentielle, loyautés contradictoires et pouvoir religieux : dilemmes éthiques dans la prise en charge du traitement de l'addiction dans un cadre confessionnel

RÉSUMÉ
Cet article fournit une plate-forme à la réflexion géographique sur les luttes invisibles auxquelles les ethnographes doivent faire face dans le domaine de la religion, de l'addiction et du traitement médicamenteux. En particulier, il examine les dilemmes éthiques et pratiques complexes impliqués dans l'ethnographie résidentielle au sein d'une communauté thérapeutique confessionnelle travaillant dans le domaine de l'addiction et de la réhabilitation. L'ethnographie

résidentielle a apporté de précieux renseignements sur la vie sociale au sein des communautés thérapeutiques et de façon plus générale, elle propose une approche éthique et participative à la recherche dans des cadres institutionnels fermés. L'immersion résidentielle dans des environnements thérapeutiques confessionnels a toutefois soulevé des défis importants concernant la gestion de l'identité, l'accès et le consentement et le dilemme des « loyautés contradictoires », un terme qui décrit un ensemble de pratiques éthiques caractérisées par un conflit éthique, un compromis et une négociation. Les chercheurs, en raison même de leur participation, sont censés se conformer à certaines valeurs, pratiques et procédés qui risquent d'être en contradiction avec leur propre éthique personnelle. Pour fixer la discussion sur la nature diverse et contestée des loyautés contradictoires, cet article examine l'exercice du pouvoir religieux et comment les ethnographes se retrouvent impliqués dans des séries de dynamiques de pouvoir, qu'ils doivent négocier, qui ne sont pas claires, sont désagréables et sont des sources potentielles d'exclusion. En illustrant les décisions difficiles que les ethnographes doivent prendre quand ils négocient les pressions pour faire respecter – ou défier – les croyances religieuses et les pratiques dans la prise en charge du traitement de l'addiction dans un cadre confessionnel, cet article invite à une réflexion critique plus importante sur les façons dont les géographes sont impliqués sur le terrain et l'engagement pratique de déontologie utilisé pour négocier les tensions d'ordre moral.

La etnografía residencial, las lealtades mixtas, y el poder religioso: dilemas éticos en el tratamiento de la adicción basado en la fe

RESUMEN

El documento proporciona una plataforma para la reflexión geográfica acerca de las dificultades ocultas que los etnógrafos enfrentan al trabajar en el área de la religión, la adicción y el tratamiento farmacológico. En concreto, se examinan los complejos dilemas éticos y prácticos involucrados en la etnografía residencial dentro de una comunidad terapéutica basada en la fe que actúa en el área de la adicción y la rehabilitación. La etnografía residencial proporcionó información valiosa sobre la vida social en la comunidad terapéutica y, en general, ofrece un enfoque ético y participativo para la investigación en marcos de instituciones cerradas. La inmersión residencial en entornos terapéuticos basados en la fe, sin embargo, planteó importantes desafíos alrededor de la gestión de la identidad; el acceso y el consentimiento; y el dilema de las 'lealtades mixtas' — un término que describe un conjunto de prácticas éticas caracterizadas por el conflicto ético, el compromiso y la negociación en los que, por la naturaleza de su participación, se espera que el investigador se ajuste a ciertos valores, prácticas y procedimientos que pueden ir en contra de su ética personal. Para asentar la discusión sobre la naturaleza abigarrada y controvertida de las lealtades mixtas, este trabajo examina el ejercicio del poder religioso y las formas en las que los etnógrafos se inscriben en, y deben negociar, una serie de dinámicas de poder que no están claras, que no son cómodas y que son potencialmente excluyentes. Al ilustrar las decisiones difíciles que los etnógrafos deben tomar en la negociación de las presiones para defender — o desafiar — las creencias y prácticas religiosas en entornos de tratamiento de adicción basados en la fe, este trabajo llama a una mayor reflexión crítica sobre las formas en las que los geógrafos están implicados en el campo y la ética práctica de compromiso utilizada para dirigir las tensiones éticas.

Introduction

Over recent years, an extensive body of work has developed to address the geographies of alcohol and drugs, addiction, treatment and recovery (DeVerteuil & Wilton, 2009; Duff, 2010; Evans, 2012; Evans, Semogas, Smallwood, & Lohfeld, 2015; Jayne, Holloway, & Valentine, 2008; Moore, 2004; Thomas, Richardson, & Cheung, 2008; A. Williams, 2016; Wilton & DeVerteuil, 2006; Wilton & Moreno, 2012). Within this, particular attention has been given to examine the ways individuals negotiate and experience different addiction treatment settings, and how specific political, racialized, gendered discourses are intimately bound up in 'recovery' (Evans, 2012; Love, Wilton, & DeVerteuil, 2012; Wilton, DeVerteuil, & Evans, 2014). While this literature is still emerging, there is a need to reflect more explicitly on the ethical, practical, and methodological dilemmas researchers face working with people recovering from alcohol and drug addiction in a variety of treatment settings.

This paper seeks to address this gap by offering a critical examination of the ethically complex negotiations ethnographers face working close-range in faith-based residential rehabilitation settings. Faith-based organisations (FBOs) are a significant feature of the broader 'recovery' landscape particularly in the area of residential rehabilitation in the UK and elsewhere (see for example, the work of Brandes, 2002; Garmany, 2010; Hansen, 2012; Neff, Shorkey, & Windsor, 2006; O'Neill, 2014; Sanchez & Nappo, 2008). Despite this, there has been relatively little sustained attention given to the ethics of engagement working inside faith-based drug and alcohol services, and FBOs more generally.

Through a case-study of an evangelical and Pentecostal Christian therapeutic community (hereafter called 'Hebron') working in the area of addiction and rehabilitation, this paper seeks to make three broad contributions. First, it explores the distinct challenges of residential ethnography, a method and ethical stance that entails living and working inside a social setting. Residential approaches to research are far from novel and hold clear synergies with other strands of the ethnographic tradition; including, overt/covert institutional and semi-institutional participant observation (Goffman, 1968; Parr, 2000); academic-activist immersion (Fuller, 1999); and 'practitioner' or 'workplace' based research (Bell & Nutt, 2002; Costley, Elliott, & Gibbs, 2010; Coy, 2006; Shaw & Lunt, 2011) whereby researchers negotiate the conflicting responsibilities working as a health, education or social care professional while conducting research in that setting. Residential ethnographies can offer a more participatory way of conducting research with people recovering from drug and alcohol addiction and potentially overcome a number of spatiotemporal barriers ethnographers commonly face in welfare organisations (DeVerteuil, 2004). However, this paper also highlights the ethical, methodological, and practical challenges involved in residential ethnography, especially in faith-based welfare settings where these issues take additional significance.

Second, the paper seeks to stimulate critical reflection on positionality and reflexivity in addiction research (following recent calls made by Jayne et al. (2008) and Lawhon, Herrick, & Daya (2014)). Geographers working in the field of drug and alcohol treatment are likely to negotiate a range of personal, professional, and academic knowledges on how 'addiction' is conceptualised and enrolled in changing regulatory landscapes and treatment modalities (Berridge, 1979; Levine, 1978; Seddon, 2010; Valverde, 1998). This paper provides a platform for geographical contributions on addiction treatment to examine more closely the ways in which specific moral, ideological, and political commitments and assumptions come to shape an ethics of engagement 'on the ground'. Following Hopkins (2007) and others (Guillemin &

Gillam, 2004), I suggest the need to be more reflexive of the practical ways in which moral standpoints and positionalities are negotiated and performed in the field, especially in scenarios in which researchers are positioned in a series of ethical entanglements that are often unclear, messy, and unpredictable. My aim is to bring to light the often undisclosed struggles ethnographers face working close-range inside faith-based rehabilitation environments.

Lastly, this paper specifically brings into focus the ethical dilemmas surrounding 'religious power' in research settings. Ethnographers working close-range in religious settings can often feel 'put on the spot' to uphold – or challenge – religious languages, beliefs, and practices of a group, or remain silent when observing conversations or practices one considers distasteful (see Chong, 2008; Han, 2010). These dilemmas take on added significance when the researcher, by nature of their religious identity, is assumed to be an 'insider' by some participants and is expected to perform set roles accordingly; for instance, by complying with the religious goals and practices of an institution. In this paper, the term mixed loyalties is used to describe a specific set of ethical practices that arise whereby the researcher, by nature of their participation, is required to conform to certain values, practices, and procedures that may contradict their own personal ethics. Through the illustration of Hebron, I present a series of vignettes that foreground the difficult and on-the-spot decisions ethnographers must make when navigating the complex ethical terrain of faith-based residential ethnography. Acknowledging the gap between textbook guidelines and ethics review boards, and the actual ethical practices in the field, this paper seeks to sharpen thinking on the ways ethnographers negotiate situations of ethical conflict and compromise researching religion and addiction in institutional contexts, and in doing so, open up a reflexive space to consider the ethics of research in these settings.

This paper is structured as follows. *Ethics and ethnography in addiction treatment settings* first situates the paper within the wider ethnographic literature on drug & alcohol treatment and institutional care settings, before introducing the case study. *Residential ethnography* unpacks the opportunities and challenges of residential ethnography, and suggests ways in which it might offer a more participatory ethic of working with residents in faith-based drug treatment services. *Negotiating religious power* grounds discussion of the mixed loyalties surrounding religious power in Hebron, and specifically reflects on the difficult negotiations I had to make as a person of faith in navigating the complex ethical and moral terrain of residential ethnography.

Ethics and ethnography in addiction treatment settings

Within geography and beyond, growing attention has been given to the ethical, methodological and practical challenges of ethnographic engagement in a range of institutional (on prisons, Drake, Earle, & Sloan, 2015; Moran, 2015; orphanages, see Disney, 2015; immigration bureaucracies and asylum detention centres, Gill, 2016) and semi-institutional settings (on mental health day centres, Parr, 2000; schools, Von Benzon, this volume; emergency and supported accommodation for homeless people and other vulnerable groups, Cloke, Cooke, Cursons, Milbourne, & Widdowfield, 2000; Cloke, May, & Johnsen, 2010; Doyle, 1999; DeVerteuil, 2004; Evans, 2012). Many of these studies highlight the ethical dilemmas researchers face adopting and negotiating different roles in the field and the distinct power dynamics these produce in relation to different groups of participants. Several scholars have discussed these dilemmas in reference to conducting participatory and ethnographic research in residential care settings, including, nursing homes for older people (Baumbusch, 2011; Tinney, 2008) and alcohol and drug rehabilitation centres (see for example, Chenhall, 2008; Mjåland, 2015; Skoll, 1992; also see Carlson et al., 2009).[1] This paper seeks to contribute to this literature and the wider geographic scholarship on spaces of drug and alcohol rehabilitation (DeVerteuil & Wilton, 2009; Evans, 2012; Love et al., 2012; Wilton et al., 2014) by reflecting on a series of practical, ethical, and methodological issues involved in residential ethnographic research in a unique type of addiction treatment setting: an abstinence-based Christian therapeutic community.

Empirical material in this paper derives from research into faith-based welfare and drug services in the UK (see Williams, 2012). By limiting discussion to Hebron, focus is given to a particular subset of faith-based residential rehabilitation providers that are characterised by a structured and overt evangelical modus operandi; and therefore do not represent the sector as a whole. Nevertheless, Hebron's network of therapeutic communities form a significant element of faith-based residential rehabilitation in the UK; offering no fewer than 340 bed places across seven cities in the UK and Republic of Ireland. Hebron is distinct in offering an abstinence-based Christian discipleship programme for men and operates independently from government funding and statutory/court referrals. The programme is optional and recommended for twelve to eighteen months during which time residents are expected to participate in daily bible study and worship as well as cleaning the house, sharing cooking rotas, and working on one of Hebron's social enterprises (gardening, renovating and selling household furniture). Prospective residents are briefed prior to arrival that they would undergo a 'cold-turkey' detoxification without any form of pain killers or substitute opiates like methadone[2] – and informed about the privations that characterised community life (no 'secular' literature, music, radio, or TV; no smoking: abstaining from sexual relations; and once-a-month family visits). For a period of two months, I conducted a residential ethnography in one of the Hebron centres, living on-site alongside 18 residents, 4 of whom had been in the community less than 4 months. The majority of residents had stayed 10 months or more, and the pastor/leader of the community lived onsite with his young family had also undergone the Hebron programme himself.[3] From the outset there are ethical questions as to whether a researcher should take up bed space in a residential rehabilitation service when currently the provision of such treatment facilities in the UK is woefully inadequate. It is important to note therefore that the residential placement took place when the group was not at full capacity and was based on an agreement that I would vacate if and when they needed the bed. During the placement I adopted the role of a guest resident/volunteer: sharing a dormitory with nine other men and participating fully in the daily structured activities, including, Bible study, prayer and worship; distributing flyers for the community business and working on the gardening and furniture renovation teams. More details about the Hebron community, including its semi-monastic culture and complex therapeutic and regulatory spaces, can be found elsewhere (see A. Williams, 2016).

Residential ethnography

Residential ethnography refers to a method of ethnography characterised by a long- or short-term residential dimension, living and working in an organisation or social setting, and adopting an overt and/or covert ethnographic stance in a group. While the merits of residential ethnographies in marginalised communities have long been established (for a recent example, see Goffman's (2015) study of a criminalised Black neighbourhood in

Philadelphia); residential ethnographic research in institutional settings can overcome numerous spatiotemporal barriers that limit researcher access, such as closing times, off-limit zones, curfews (as illustrated in DeVerteuil's (2004) reflections working in a homeless women's night shelter). As a method of ethnography, residential placements enable a more dynamic picture of the variegated spaces and temporalities that make up organisational welfare spaces. For instance, residential presence in Hebron provided insight into the ways different spaces facilitated and/or constrained particular interactions and types of conversation among residents ('recovery talk', 'spiritual talk', 'subversive chat' and 'downtime'). Chatting in the dorms with residents or delivering flyers advertising Hebron's social enterprise meant levels of supervision from leaders and other residents varied considerably, which in turn resulted in greater opportunities for residents to speak 'freely' (although it is important to acknowledge conversations were always tied to shifting social dynamics between residents, alongside their perception of myself as an 'outside researcher'). This allowed me to be privy to often fleeting comments made by residents and staff as well as access to the daily rhythms and unwritten rules of an organisation, including observation of the 'darker side' of organisations that might remain hidden from shorter-term or transitory research engagements. Residential ethnography in this way is uniquely placed to 'analyse hidden agendas, 'taken for granted' assumptions ... working towards disrupting the status quo by bringing into light systems of power and control (Baumbusch, 2011, p. 185). In so doing, it raises complex power dynamics for researchers observing practices and relationships spokespeople of organisations might either be unaware of or want to conceal (see, for example, the dilemma of whistleblowing in Greener's (2015) covert ethnography in residential care settings for older people).

Residential ethnography can be understood also as an ethical stance that seeks to create participatory routes for individuals to 'speak out' and 'speak back' to the researcher. Following Parr's (2003) call for research to include the perspective of those who are 'cared for' not just those that do the 'caring', residential ethnography in treatment settings represents an approach that seeks to access the experiences of those recovering from alcohol and drug problems, who often find their voices silenced, unheard and frequently distorted in popular accounts, including the publicity campaigns of charitable organisations working in addiction treatment. Residential ethnography was particularly suited to revealing the different ways individuals gave meaning to and experienced drug and alcohol treatment organisations, especially in closed structured programmes which can have few avenues for service-users to communicate criticism. Living onsite not only allowed greater opportunity for trust and rapport to develop with participants, but also the ability to move between various 'private' and 'public' spaces in the community to facilitate confidential conversations where residents could speak without fear of reprisal. Residential ethnography also created space for participants to discuss and challenge researchers' interpretations in the field and to reflect on their own immersion in the social setting. Practically, this entailed sharing preliminary interpretations in a way that allowed participants to reflect, correct and challenge, while, at the same time, being sensitive to the malleable power-dynamics and unwritten codes this might reveal.

Residential ethnography can therefore allow a degree of experimentation in the field as researchers immersed in the physical and emotional activities of an organisation develop strategies to encourage participants to reflect on the moral codes, atmospheres, and structures that make up organisations. Living and working inside faith-based drug services, especially those which mandate participation in religious practices as part of their programme, allowed me to gather a more dynamic and embodied understanding of the different ways individuals came to narrate, negotiate, perform, experience and contest religious space, identity, and belief. Over time, residents offered to show letters and diaries they had kept since arriving at Hebron. Discussing these materials together gave access to the different ways individuals ascribed meaning to, and experienced, the day-to-day routine in Hebron and how these changed over time. The two-month placement in Hebron, albeit a relatively short period of time, allowed a degree of friendship to develop with residents through the comradery on the work teams, by teaching new songs on the guitar, and engaging in private conversations with residents about their faith, as requested. In a small way, this gave me opportunities to meaningfully give 'something back' beyond some vague allusion to bettering academic and policy-makers' understanding of faith-based residential rehabilitation.

Such an approach to residential ethnography afforded a more participatory ethic to research with a particular vulnerable population; but it raised a number of practical, ethical and emotional challenges to which we now turn.

Negotiating access

The Hebron community in which I stayed operated on a shoestring budget and was largely reliant on public donations and income from its social enterprise. Unsurprisingly the leaders were initially wary of the damaging representations an outsider might paint of the community and commented that a faith-based charity had recently lost government funding in claims that it used public money to 'proselytise' (see also Davies, 2004). Guarantees of confidentiality – with regard to the use of pseudonyms to protect the identity of the organisation, its location and names of all participants – played a part in gaining consent from organisational gatekeepers, but more important was the deployment of my own Christian positionality in negotiating access to Hebron and other 'closed' evangelical Christian groups in my research. Even so, on arrival in Hebron I had to earn the right to speak by passing through a series of legitimacy and belief tests as staff and residents determined the motives underpinning my research and how the research was going to be used. While the trust gained from performing a degree of cultural competence (Mohammad, 2001) helped reassure those hesitant to open their doors to the possible 'misinterpretations' of social scientists; this in turn raised questions about the politics of disclosure and the practical ways a religious 'insider/outsider' communicates more critical findings about an organisation without it constituting a betrayal of trust. To navigate this, the research project was framed around 'the role faith plays in faith-based drug services', emphasising this was not about a positive validation of faith-based activity in welfare, but a search to develop more grounded understandings of the day-to-day theologies and ethics of care performed and experienced by staff, volunteers and service-users in faith-based drug and alcohol treatment. Using the lexicon of 'good practice' and 'service-user experience', I tried to distance myself from Hebron leadership and emphasise to each participant that my interest was in their own experience and personal account of Hebron, whether positive or negative.

However, it soon became apparent that negotiating access in Hebron was not a 'one-time hurdle'; but a dynamic process that was negotiated each day and in relationship to different groups of participants (Crowhurst, 2013). Accessing prospective interviewees and gaining 'informed consent' was more complicated as a guest in a closed institutional settings and I relied on gatekeepers for their continued goodwill and cooperation. As a result, I largely felt compelled to work within the discursive norms and codes of Hebron's organisational culture given that withdrawal of leaders' backing for any reason might signify a premature end to the research.

Problematising consent

The second challenge I faced working in Hebron was the hierarchical nature of organisational decision-making. After arriving, interview access to 'new' residents was restricted by staff, reneging on a prior agreement. Instead I was only permitted to 'formally' interview those who had been at Hebron for over two years and supposedly would give a 'good account of the centre'. The two leaders, Matt and Liam, did not want the interviews to disrupt residents' work schedules, and cautioned that any participation in research might have 'negative influences on the guys who had only been [t]here a few weeks or 2 months' (interview with Liam, 25/04/10). Given that I was sharing a dormitory and spending the work day with new residents this interview restriction was not an impediment to accessing individual stories and experiences of Hebron. However, this situation raised serious questions about the ethics of covertly documenting informal conversations in an organisation that has retracted the freedom to interview some participants under the presumption this might be detrimental to their engagement with the treatment programme. It also highlights the problematic of 'coerced consent' whereby an organisational gatekeeper can grant (or withhold) permission to speak to 'service-users' who subsequently might feel obliged to offer a positive portrayal of the organisation (or keep silent). More subtle forms of coerced consent often go unacknowledged in welfare settings linked to the distribution of goods (see, for example, a needle-exchange, food bank or night-shelter; see Bourgois, 1998; Garthwaite, 2016; Moore, 2004). In these cases, prospective research participants might feel obliged to respond positively to the request for interview to ensure receipt of those goods and/or rehearse narratives of gratitude. The research encounter is shaped by broader power-dynamics that risk reproducing a subject-position of disempowerment and disenfranchisement. The politics of fieldwork can therefore work to exacerbate, rather than ameliorate, stigma of vulnerable groups in welfare settings.

Emotions in the field

There are several ways in which residential ethnography makes the researcher vulnerable in ways in that other ethnographies do not. First, the residential placement gave minimal space for respite or escapism during my research in Hebron. The effort put into 'earning' trust and building rapport in the community meant I constantly felt 'front-stage', aware that others might be scrutinising my performance. Navigating, and/or conforming to, the expected values and practices of a closed group can be a significant source of emotional conflict for the residential ethnographer (Chong, 2008). To be welcomed and accepted in a sensitive, close-knit environment, researchers require a range of social interaction skills and a willingness to participate in all activities. In the case of Hebron, this involved fully immersing myself into the daily routine and 'pulling my weight' on the work teams, with their distinct macho-culture that jokingly tried to 'make a man out of this pen-pusher [me]' (José, resident 13/05/10). Alongside the emotional toll of identity management in the continued negotiation of research access, there was limited opportunity to offload or debrief with friends or family because access to my phone had been restricted to 'remove any temptation from the lads' (interview with Liam). Working the same hours as residents meant there was limited designated time for 'formal' interviews. Neither was there sufficient time to process research notes as the structured routine of the community entailed only one hour of free time which I felt compelled to spend in the lounge with other residents.

Second, residential ethnography makes the researcher vulnerable to assimilation into the social and affective rhythms of the group because it entails a total, albeit time-limited, immersion into an institution. While there is a danger that assimilation into the religious culture might compromise the research, becoming too familiar with its rituals and performances as to lose critical edge; the gradual edging towards 'going native' is analytically interesting for that very reason as it provides an unparalleled insight into the micro-politics of lived religion in faith-based addiction services. When such experiences were shared in dialogue with participants in Hebron it helped foster critical reflection on the ways Hebron residents came to perceive, narrate and perform religious spaces, beliefs and practices.

Lastly, deconstructing one's own faith and that of others can be a deeply uncomfortable and vulnerable position; one that simultaneously brings emotional upheaval, disenchantment or disconnection, as well as spiritual depth and renewal. As friendships developed with Hebron residents, I felt increasingly torn about the purpose of my research, and indeed the ethics of challenging or undermining the efficacy of transformational narratives of conversion and deliverance so many residents passionately shared with me. While I remained critical of any consequentialist ethic that implies the 'ends justify the means', it did push me to write an account of these spaces that sought to portray the complex intersections between therapeutic, spiritual, emotional, and regulatory experiences. Even as I write this paper, I am conscious the research experience has instilled an analytical sensibility towards the curation and performativity of worship space that at times has resulted in estrangement from elements of faith practice previously I felt comfortable with. This underlines the importance of critical reflection on politics of representation, a topic to which I return below.

Mixed loyalties

Finally, residential ethnography raises serious questions about how researchers should engage in the ethically complex spaces of faith-based residential rehabilitation. The nature of my involvement in Hebron immersed me in a web of power-relationships between different groups of staff and residents, each of whom had specific expectations. The term mixed loyalties[4] is useful here to foreground the varied and ambivalent situations of ethical conflict, compromise and negotiation whereby the researcher's affinity or duty to a particular set of organisational rules, roles, and relationships is called into question. As discussed previously the term has specific credence in welfare settings where the researcher, in performing their role as a volunteer or staff member, is required to follow particular procedures that may contradict their own personal ethics; for example, the ethical dilemmas entailed in complying – or not – to 'referral only' policies in emergency accommodation for homeless people, or the 'no voucher no service' rule in many UK food banks (Williams et al., 2016). Attending to mixed loyalties in research encourages critical reflection on the unclear, uncomfortable, and potentially exclusionary situations whereby the positioning of the researcher in relation to other participants (staff, volunteers, service-users) involves the exercise (or passive observation) of varying degrees of social control, discipline and surveillance. Conceptualising mixed loyalties in Hebron, however, is not only confined to the quandary of organisational compliance. Rather, in religious settings a distinct set of mixed loyalties can arise if the researcher, who identifies as a person of faith, is assumed to be an 'insider' and expected to uphold the religious values and practices of the community.

The remaining section of this paper reflects on some of the significant ethical dilemmas that arose in the residential ethnography of Hebron. Particular attention is given to the ethically complex exercise of 'religious power' as a way of illustrating the need for greater analytical reflection on the practical and ethical dilemmas researchers experience working inside faith-based drug treatment settings, and religious environments more broadly. While this paper in no way seeks to provide a comprehensive theorisation of religious power, some conceptual acknowledgement of religious power and its contextualisation within Hebron is necessary.

Negotiating religious power

Recent scholarship on religious power[5] has increasingly drawn on Foucault's later writings to examine the intersection of religion, the body and power (Carrette, 2000, 2013; Holloway & Valins, 2002; Garmany, 2010). Building on rich accounts of therapeutic and regulatory power in addiction treatment (Wilton & DeVerteuil, 2006), the construction of subjectivity in harm reduction programmes (Fraser & Valentine, 2008), and the ethical problematisation of the self in controlled drinking units (Evans, 2012); I wish to highlight two aspects of religious power that will help contextualise the emergent ethical dilemmas and power-dynamics in Hebron. The first refers to the religious problematisation of addiction and the disciplined body: namely, how organisational logics, architectures, and daily routines were shaped by a particular religious set of assumptions about the nature of, and solution to, 'addiction' (cf. Wilton & DeVerteuil, 2006). Specifically, this refers to the belief in Hebron that the source of addictive behaviour was rooted in sinful desire and its trappings of shame and guilt. Religious conversion, understood as accepting Christ's unconditional love and gift of righteousness, was presented as a means to bring radical and long-lasting personal change that broke the bondage of addiction. This belief underpinned pastoral power in the community, premised on submission to religious instruction through mandatory participation in Bible study and worship; alongside the peer expertise of 'former addicts' and an explicit suspicion of self-help/medical/psychiatric interventions. Intimately bound up in this were embodied relations of religious power, manifest, for example, in the way in which privations were rationalised (including the denial of psychoactive medication, smoking, and the prohibition of 'non-Christian literature') and codes of etiquette and interaction (or 'participatory manners' in worship space) were oriented by religious instruction (A. Williams, 2016).

Second, while it is clearly important to emphasise the highly regimented and controlling characteristics of religious power in Hebron, it is equally important to acknowledge that Foucault's own writings on spirituality and religion, especially in his later works on sexuality, ethics and subjectivity (Foucault, 2001), moved beyond accounts of disciplinary, repressive or self-regulatory power, and provide instead tools to study the more 'productive' nature of religious power in relation to technologies of the self and ethical problematisation. In this way, a more dynamic conceptualisation of religious power emerges through analysing the nexus of governmental and ethical processes of self-formation, examining the ways in which religion delineates particular types of attitude, desire, and behaviour; but also how bodies use religious discourse and practice to mediate and resist power (see Jordon, 2014). Religious practices in Hebron therefore relied on and worked through residents' own ethical problematisation of the 'addicted self' (and a range of other mediatory 'truths'[6]), whereby individuals confronted the self as an ethical problem and sought to transform and modify themselves accordingly.[7] Such a perspective helps foreground the complex interplay between 'top-down' religious discourses and practices, and the lived negotiations of residents, their own cultural competencies, and strategies and tactics used to 'question, form, know, decipher and act on themselves' (cf. Dean, 1995, p. 563).

Significant questions remain as to how researchers should negotiate religious power, especially in faith-based treatment settings where the researcher might feel compelled to work within the discursive norms and codes of the organisation to maintain gatekeeper access. Through a series of vignettes drawn from the residential ethnography, I examine the distinct ethical dilemmas facing ethnographers working inside closed religious environments. Particular attention is given to situations of ethical ambivalence and conflict resulting from my own engagement with 'religious power' inside Hebron, including: the use of scripture; performance in worship services; adopting a position of confidant over matters of belief; and observing forms of bullying and the temporary denial of medical attention.

Mixed loyalties and religious power

Four weeks into my placement in Hebron I witnessed a new resident being 'teased' about his weight by two more established residents. Four of us were working on the garden team and were behind schedule relaying a lawn. Liam, the deputy leader of the community, was gradually getting more frustrated with Karl, a new resident who apparently 'wasn't pulling his weight' and regularly needed to sit down. Karl had had a conversion experience in a Pentecostal church several years ago and was anxiously asking whether he was 'still a Christian after relapsing'. Liam, without raising his voice, said firmly:

> Your brain is all gunked up with 'weed'[cannabis] ... It will take a while for it to leave your system, maybe two months. (ethnographic field notes 20/04/10)

This paternalist construction of the 'irrational addict' led to rather blunt forms of encouragement:

> It was your choice to go back on drugs – you turned your back on Jesus and all what he has done for you. Once you realise what God has done for you, deep down, you wouldn't want to go back to drink or drugs. Christ gives freedom. (Liam, ethnographic fieldnotes, 20/04/10)

Karl remained silent and on appearance did not seem uncomfortable with what Liam had said. I was initially shocked on observing this moment, frustrated that this view went supposedly unchallenged. I tried to interrupt the conversation by using a theological script I thought would be accepted in the community to question the way religion was being enrolled into such a stigmatising narrative:

> Well, it's complicated. We will all go through struggles, being a Christian doesn't make us immune from that... Even in the Bible,[Saint] Paul talks about getting stuck in a rut in 'not doing the good he wants to do, but doing the things he doesn't want to do'. I think we can all relate to that, whether it is drug addiction, lust, greed, pride, take your pick, we all mess up, one is not worse than the other. But God is like that father in the story of the Prodigal Son, each night he waits for his son to return. He has not washed his hands with us when we sin but waiting to welcome us back. (ethnographic fieldnotes, 20/04/10)

This extract illustrates the 'on-the-spot' ethical dilemmas researchers often face when working with 'vulnerable' adults in organisational spaces of care, and raises a specific set of questions for geographers of religion involved in 'close-up' research inside religious communities. To what extent, and in what ways should researchers utilise the religious language, practices and norms of the community when engaging individual participants? For instance, my choice to present an explicit Christian message depending on one's vantage point could be regarded as oppressive: not challenging head-on the overarching narrative of addiction as 'moral failure' and enculturating instead a particular religious relation of self to self, premised on guilt (sin) and need for release from a Big Other (salvation). The rationale for using a theological script (as opposed to more secular academic/practitioner language) was threefold. First, it was based on an understanding of Karl's own faith-position and the desire to be responsive in a way that was meaningful to him. Second, perhaps pragmatically, it stemmed from the perceived need to speak the 'right language' so as to not 'rock the boat', both in terms of the individual well-being of participants, and the need to retain sufficient rapport and trust with leaders. Lastly, as a practicing Christian, there was a greater personal stake in challenging what I saw as a politics of fear instilling feelings of personal failure that, in some cases, induce behaviour amenable to manipulation and control. My own personal faith journey in and out of various elements of charismatic and evangelical Christianity has over time given me a heightened sensitivity both to the dynamics of religious power (as someone practically involved in pastoral care with church members struggling with alcoholism and chronic mental health difficulties) and the dangers of guilt-inducing narratives that can pervade, sometimes inadvertently, forms of religious instruction. For this reason, I was critical of the modus operandi of Hebron and similar programmes premised on mandatory religious participation and geared towards reforming the habits and desires of an individual towards an 'idealised 'Christian identity.

Yet while I disagreed with Hebron's ideology and practices, I found my position as a volunteer becoming increasingly pragmatic: prioritising ways in which I could aid personal well-being and motivation in residents' recovery. Despite my familiarity with academic criticisms of different medical, social and spiritual approaches to substance misuse and addiction, in the moment my own ethics of engagement were as much shaped by personal experience and situational encounters in the field as by these academic knowledges. Personal standpoints on 'addiction' and 'drug use' invariably change over time – and will continue to change in relation to a multiplicity of different events and one's relationship to these events. At the time of fieldwork, my assumptions on drug use had been largely overshadowed by a friend's untimely death to an ecstasy overdose, alongside a series of experiences volunteering in homeless shelters and daycentres: hearing first-hand accounts of street drug use and overdose; facing violent demands for money and threats of self-harm; and being called to administer emergency first aid to a service-user who had hit an artery 'shooting up' in the toilet and had passed out. There is no easy translation between trauma and the personal, professional, and political standpoints academics take on substance use. For me, these experiences have led to a broadly defined recovery-focused approach[8] in my academic and voluntary work holding rather pragmatic views on abstinence and harm reduction debates yet remaining critical of the criminalisation of drug use.[9] In turn, I have tried to embody a non-judgemental and emotionally responsive stance that works with the beliefs and resources of the individual.

Engaging with residents who were keen to share their stories of recovery and religious conversion put me in an uncomfortable position as to whether or not I should challenge what I saw as a highly stigmatised construction of 'the addict' alongside overt criticisms of alternative ('secular') approaches such as harm reduction. Early on in the placement I was keen to earn trust and not explicitly disrupt the dominant moral and religious discourses in the community, in case it was detrimental to an individual's recovery. I was also hesitant not to be perceived as bringing my own middle-class respectabilities or 'elite' education to a therapeutic community where staff and residents all had personal experience of chronic substance misuse. Over time I became more comfortable raising questions about the limits of Hebron's theological construction of addiction (with its emphasis on conversion and detachment from the world), and the ways Biblical text was sometimes invoked to shut down debate. Engaging religious power, particularly via the use of scripture in everyday conversations produced at times an uncomfortable dynamic with leaders and more established residents which, in some cases, was perceived as a threat to their interpretation of scripture, and by inference, their 'recovery' authority.

In specific situations where residents approached me to discuss personal and theological issues, I first sought to discern and engage residents' individual frameworks of belief, identity, and spirituality. I did not have a blanket rule about not 'doing God', but I deliberately avoided religious language when speaking to new residents and/or those who identified as non-religious. More generally, I was wary of the unseen implications surrounding my use of religious language. For some residents it potentially ran the risk of further legitimising a discursive environment in which 'faith' is held up as a panacea for addiction and complex social and psychological conditions (see Williams, 2013). Perception of the ways I talked about Christianity in conversations, combined with the affective manner in which I performed (or not) religious practices such as prayer and worship opened up – and closed down – different sets of relationships with groups of residents: some regarding my participation in worship services as a sign of congruity, trust and safety ('one of us'), while others seeing my involvement as a form of estrangement ('one of them'), even perpetuating a sense of isolation. Yet throughout the placement I was explicit in distancing my own faith identity from that of Hebron's, and in one-to-one and group conversations made sure I was open about my reservations about aspects of Hebron in hope that this might empower some residents to voice their views also.

'Taking sides'?

Over the course of the placement I became particularly close to several residents who at times used me to emotionally offload and question the validity of leadership decisions. Karl was missing his young family back in the Midlands and came to me saying he was 'feeling fine and ready to go back'. He had already packed his bags ready to leave. Faced with his uncomfortable dilemma I suggested he best think it though and speak to others in the community. He was talked out of leaving by Liam. During the worship session the next morning, Liam announced:

> Leaving after a few months is a waste of time. You may think you're fine and free from addiction but you've still got loads of stuff you need to deal with. If you want to change you must be willing for real discipline and structure... Don't treat the eighteen months as a jail sentence. If you think about it in the same way you will leave here unchanged. It will work if you work it, being successful here is showing and taking responsibility: 'stepping up to the plate'. (ethnographic fieldnotes, 04/05/10)

On the way to a garden job later that day, Liam explained 'during the first few months you've got to push the new guys coming off addiction, otherwise they will fall when their bodies get ill; they won't change'. The somewhat uncomfortable positioning of confidant with some of the residents put me in a difficult ethical position: what do you do when someone expresses desire to leave the community? It is in these moments researchers are perceived to 'take sides' with the community or the individual. By not 'getting involved' initially and regarding Karl's wish to leave as 'just a low point', I ensured I did not step out of line with others in the community. In the days that followed I tried to explore some of the deeper reasons why he wanted to leave but at the time he did not want to talk about it. Instead, Karl seemed to have developed a renewed faith and on at least one occasion started preaching to other residents such as Paul: 'if you have faith you cannot be addicted because Christ lives in you'. 'But' Paul replied 'but do you have to already have faith to have faith?' 'That's the devil attacking me' Karl stated, pointing at Paul: '[it's the Devil] using people to confuse me. You don't understand because you're an atheist'. Paul hit back in frustration: 'you've just got a simplistic narrow view of life; anything outside your belief system you see as wrong. If I don't adhere to the way you believe then I am called an atheist. I am not an atheist!' This example demonstrates that the dilemma of 'taking sides' was not limited to a staff-resident dynamic, and raises questions as to how researchers should position themselves in situations of religious disagreement between residents themselves; and whether or not, and to what ends, should the researcher challenge the unwritten and hidden assumptions in closed communities.

The challenge of 'taking-sides' among different groups of residents and staff became most public when Karl requested to see a nurse. At first Liam refused, stating unless it was urgent Karl would have to show his commitment to Hebron by staying three months: 'Otherwise we have people coming here, get treatment, which is very expensive, and then leave'. Karl had been at Hebron two weeks and explained to the group he had occasional minor bleeding from his rectum and was in pain when he moved about. Following a more sympathetic decision later that afternoon from Matt, the senior leader, Karl received medical care within two days, demonstrating both the discretionary power held by leaders, even to the extent of withholding medical treatment.

In Karl's case, other residents had conveyed concern and understanding of his situation. However, this was not always the case. A more established resident, Graham, in his fifties had a reputation among the group for 'crying wolf'. He had sores over his feet and after obtaining medical attention had to go to the nurse once a week to get a bandage dressed. Despite his doctor advising him of the need for rest and not to put any pressure on it, this request was not taken into account by other residents who suggested 'he's faking it just to get off hard work' (ethnographic fieldnotes, 15/05/10). On inspection Graham appeared to be doing more damage to his foot. It was a week before the doctor's advice was finally taken into account. Graham conveyed his sense of disempowerment and frustration in front of a small group of residents:

> That's just the way it is ... maybe I can go to another rehab, give it eight weeks rest and it will heal... But they [Matt and Liam] 'know best...' (ethnographic fieldnotes, 15/05/10)

In Graham's case, I was caught up not simply in a staff-resident hierarchy, but in a more complicated set of personal relationships between different groups of residents. I tried to avoid 'taking sides' initially but as the evening went on I pressed some of the residents to believe Graham's case. However, it is equally as important, if not more so, to address instances where I chose not to 'speak out' and in doing so implicated myself in tacit tolerance for disciplinary practices. Halfway through my placement Alex, a new resident of four weeks, was caught smoking a cigarette he had picked from the pavement; he decided to leave after he was put on dishes for a month as punishment. Nikolay, a 26-year-old resident had been in the community four months and was put on dishes for two months for shaving his head without permission. 'It makes you look like a thug and we can't have that if you are on pick-up and delivery [of furniture]' Liam stated. In these situations, my role of a guest/volunteer meant it was 'not my place' to challenge such decisions. On another occasion, Paul, a Hebron resident of three months expressed the desire to leave the community when we were out of earshot of other residents:

> Fuck this, I just want out, I am fed up with people telling me what to do, the shit unquestionable authoritarian structure, I miss having freedom, I miss my kids...

Paul was in his mid-fifties and had previously been a physics teacher in a boys' grammar school, turning to alcohol to deal with stress of performance indicators. This was the first time he had shared in such strong way his frustration with Hebron. During previous conversations I had already distanced myself from Hebron theology and we had bonded over a mutual dry sense of humour. Knowing that we could be interrupted whilst talking and vacuuming the dormitories, I tried to voice some of my concerns about Hebron whilst at the same time attempting to give some assurance that 'things will get better' – knowing that encouraging him to leave would likely have led to my swift departure also. It was also in part a sign of my own immersion in Hebron's recovery language that considered the mind-body-spirit as 'vulnerable' to relapse. While the residential nature of the fieldwork allowed deeper follow-up conversations about Paul's experiences, the immediate impulse to reassure, comfort, or care rather than ask critical questions, might have been read by Paul as not being concerned about or validating his experience – giving the impression I was either not actually listening, or worse, choosing siding with the leadership and encouraging him to simply stay with 'the programme'. Analysis of these ethically important moments (Guillemin & Gillam, 2004) illuminates both the blind-spots inherent in the ethnographers' gaze and the ways these can shape discernment of, and choice of action within, ethical entanglements; as well as highlight the possible constraints researchers face trying to instigate change in hierarchical power structures.

Vulnerability, proselytisation, and the politics of representation

The above vignettes have sought to give an honest, albeit vulnerable account, of the practical and ethical dilemmas and compromises I faced working in a faith-based residential rehabilitation setting. Working as a person of faith inside a Christian therapeutic community raises significant questions as to how my own faith-based positionality came to shape my view of 'vulnerability' of Hebron residents and the politics of representing the personal stories and organisational practices of Hebron. My preliminary interpretative framework prior to entering Hebron had been informed in part by a critical curiosity towards the triumphalist accounts of miraculous healing and deliverance from addiction found within evangelical Christianity (see, e.g. Pullinger, 1980; Wilkerson, 1963) alongside contrasting accounts of indoctrination, abuse, and control in prominent Christian rehabs in the USA and elsewhere (see Pollard, 2008). Holding these narratives in tension pushed me to examine the 'agency' of lived religious subjectivities practiced and experienced by (some) residents, and contextualise this within the complex therapeutic, regulatory and emotional geographies that coproduce life in Hebron. 'Vulnerability' in Hebron can be conceived on several registers: (i) the lack of privacy and separation from families and loved ones; (ii) limited contact with qualified medical and psychiatric care could exacerbate, rather than alleviate, mental health symptoms; (iii) concerns about non-medicated detoxification and withdrawal (especially for alcohol), which in some cases, can induce psychosis and hallucinations; (iv) the elevation of 'faith in Christ' as a positive replacement for 'addiction', combined with restrictive autonomy, can be viewed to exacerbate forms of co-dependency whereby residents become institutionalised into a religious lifestyle and trapped within and dependent on the regimented structures of community; (v) Hebron's prohibition of any opiate-based or psychoactive medication (including anti-depressants) might mean some mental health issues may go untreated, or remain undiagnosed or misunderstood by others; (vi) cases of bullying and emotional coercion experienced by residents; (vii) lack of aftercare facilities on leaving a regimented environment; and (viii) unwanted/unethical forms of proselytisation that exploit the power imbalance between service-provider and user.

Here there are critical questions about 'volitional agency' with regard to the intersection of social attachment, trauma, and the overtly regimented exhortation to a particular religious belief and discipleship (Williams, 2013). While several faith-based drug and alcohol programmes in the UK have, according to a recent Demos report, made a 'shift from proselytising and making services contingent on compulsory attendance of religious services' (Birdwell, 2013, p. 36), Hebron eschews public funding to maintain a distinct evangelical modus operandi: offering a Christian discipleship programme that is voluntary in nature and claims to be explicit in explaining to prospective residents (and their families) the religious expectations and semi-monastic structure of the community. Despite several residents consciously choosing Hebron because of its faith-based character; for others, who were encouraged by family, recruited in homeless shelters, on the street, or through informal referral routes with probation officers, there are important questions about 'informed consent'. Such recruitment strategies into a Christian discipleship programme might be read by a sceptic as simply a mechanism that preys on vulnerable people who might be desperate to change, and ready, if at times hesitant, to accept a promise of salvation, sobriety and a new life. Certainly, the notion of 'freedom' underpinning the optional nature of Hebron ('free entry and exit') needs to be deconstructed given notions of 'choice' are inherently caught up in complex social attachments within and beyond the Hebron community, and that in some cases, residents lacked viable alternatives if they left Hebron.

While I remain troubled by theological, cultural and social characteristics of this particular model of semi-monastic therapeutic community, it is important not to treat its residents as simply passive recipients or religious dupes of a structured regime. Nor should we typecast Hebron staff was mere disciplinarians and ignore their sincere commitment to spend much of their lives living *with* and caring *for* a marginal and highly stigmatised population, and in doing so, taking vows of material poverty, shared finance, and chastity (for non-married members).[10] As I have argued elsewhere (A. Williams, 2016) interpretations of such environments that only address technologies of control overlook the other logics and processes (compassion, friendship, spiritual, therapeutic) that coexist in these spaces, and understate the ways staff and residents are 'fully fleshed' subjects in their own right, who bring their own strategies and tactics to engage (or not) in religious therapeutic spaces. Proselytisation in Hebron therefore occupies a deeply ambiguous and ethically contested space, one that requires grounded analysis of the variants of physical, psychological or social coercion *alongside* the ethical agency and capabilities of individual residents, many of whom brought strong existing religious and spiritual beliefs to Hebron. Indeed, Hebron residents who believed in the therapeutic and transformative potentialities of Christian therapeutic community did not see themselves as 'vulnerable' to indoctrination and contested such a labelling when the issue was broached. Vulnerability instead was framed in relation to the need to 'protect' their own recovery and that of others from the risk of relapse. For others, however, there was an explicit criticism of Hebron's strict regime and the effectiveness of a faith-based approach to addiction treatment. Any assessment of these ethically complex spaces must therefore take seriously the lived subjectivities of residents and the ways in which individuals experience, perform, and contest the variegated emotional, regulatory, and spiritual geographies that co-constitute social life in faith-based residential rehabilitation environments.

Conclusion

My purpose in this paper has been to highlight the hidden struggles researchers can experience conducting residential ethnographies in faith-based addiction treatment and rehabilitation settings. Via the illustration of Hebron, this paper has developed a series of insights into the ways that geographers might approach residential ethnography in drug treatment and other care settings. The merit of residential ethnography is threefold. First, as a methodological approach it provides rare access into the sometimes concealed realities of residential care institutions, and in doing so, a more variegated understanding of the different spatialities and temporalities that make up organisational spaces of care. Second, residential ethnography can offer a pathway for ethical and participatory research in closed settings – one that potentially gives space for residents to 'speak out' about latent concerns as well as 'speak back' to the researcher in ways that help refine research objectives and provisional interpretations. Although not fully advanced in this paper, this process also includes giving space to hear the ethical deliberations of leaders who might be dismissively portrayed as 'discipliners' to residents; and instead open discussion of their own 'mixed loyalties' (negotiating the personal vs procedural) and the entangled organisational, therapeutic, religious and emotional commitments at work in decision-making. Third, residential ethnography as a volunteer denotes a 'participatory research practice that enables the researcher to actively contribute to constituting and reproducing an organisation. The research becomes a way to 'excavate, experiment, evaluate [and] amplify alternative trajectories (Iveson, 2010, p. 439) through participation' (cited in M. Williams, 2016, p. 2). Residential participation can facilitate important opportunities for influencing change in organisational practices, as well as bring to light the 'darker' aspects of care/welfare spaces. However, residential ethnography is not without significant challenges, including, negotiating access and consent; identity management and emotional involvement; alongside the deeply ambiguous, uncomfortable, and exclusionary relations of power residential ethnographers can become enrolled in by nature of their participation. Specifically, this paper has focused on the everyday negotiation of 'mixed loyalties' in relation to religious power as means of highlighting the ambiguous and conflictual decisions ethnographers face working in closed religious environments, but also welfare organisational settings more generally. Through this, I hope reflection on these real-life research entanglements will have broader pedagogic value and give greater space for other researchers to acknowledge the hidden struggles and power-dynamics that emerge in a range of different religious, welfare, and addiction treatment research arenas.

In all, this paper raises a series of questions that have wider implications for geographic research in the areas of religion, welfare, and addiction treatment. First, for geographers working in the field of substance use and addiction treatment, this paper calls for a more explicit reflexivity on the moral and ideological assumptions researchers make about the nature of 'addiction' and 'treatment', and how these shape the discernment of, and the process of engaging within, a range of ethical dilemmas that arise in the field. Researchers will occupy a range of different standpoints on abstinence/harm reduction debates; medical, social and spiritual models of 'addiction'; criminalisation/legalisation; regulation/deregulation, and it is important to acknowledge the complex and shifting intersections some scholars take across these vectors of debate. In what ways, and why, do political commitments and moral standpoints shape (i) the types of organisations, spaces and participants studied; (ii) the assumptions made about the 'vulnerability' of individuals; and (iii) how do these views come to shape an ethics of engagement in the field? These standpoints will have an enormous bearing on the interpretation of drug and alcohol treatment settings, what we prioritise in our writings, and the manner in which we negotiate and perform ethically complex moments that arise with participants in the field.

Second, the paper questions the extent to which geographers of religion actively reflect on the dilemmas of 'religious power' and its implications in generating ethically ambiguous sets of power-dynamics in research. Careful reflection is needed as to how researchers negotiate questions of intimacy and collusion when working close range in therapeutic and/or religious environments; especially when the researcher/volunteer occupies a position of friendship, responsibility or therapeutic authority. If the hidden stories of religious ethnography are left untold, then geographical scholarship on faith-based reflexivity risks remaining stuck in well-worn debates about the merits of 'insider' and 'outsider' researchers, rather than examining the analytically more interesting practical ethics and struggles ethnographers of religion embody and perform in the field. While this paper has largely focused on the ethical dilemmas of religious power in faith-based residential rehabilitation settings; the discussion here is relevant for thinking about the hidden struggles ethnographers face working in other arenas of substance use, harm reduction, and addiction treatment, and encourages critical reflection on the practical ways researchers engage or negotiate the goals, languages, and practices of dominant treatment philosophies and regimes, especially those which the researcher might find problematic or distasteful.

Lastly, and related, to what extent and how do researchers reflect upon and evaluate 'mixed loyalties' in research? Mixed loyalties are situational and in a large part shaped by the positionality and reflexivity of the researcher and the nature of their involvement in the field. While it is important to reflect on the mechanisms by which the more negative experience of personal conflict and ethical compromise are negotiated and actively managed, it is vitally important to maintain a more productive reading of 'mixed loyalties' in research. As an ethical practice that a researcher purposively adopts in research, being reflexive of 'mixed loyalties' in the field can open up more experimental lines of engagement that expose and cut against the dominant political and moral logics in different settings, and through grounded participation offer routes to generate reflexive dialogue among participants when working from positions of more-or-less explicit opposition (Thiem & Robertson, 2010). It is in exploring and actively working the ambivalent spaces of mixed loyalties, as opposed to simply bypassing or seeking to resolve these tensions that helps illuminate the hidden and unwritten codes, exclusions and power-dynamics in mundane spaces of care.

Notes

1. Ethnographic and participatory work with street drug cultures present a series of additional ethical, methodological and practical challenges. Many of these issues have been addressed in Sandberg and Copes' (2013) excellent account of the 'hidden ethnographies' of street drug cultures, including ethical and practical dilemmas ethnographers must negotiate in the field, with regard to, for example, drug participation, 'taking sides', informed consent, confidentiality, payment, and ensuring physical safety and legal security in the field for both participants and researchers.
2. Individuals with higher support needs were encouraged to seek medical detoxification or psychiatric support prior to arrival.
3. It is common in evangelical Christian drug rehabilitation programmes for leaders and staff members to be former 'graduates' of the programme in order to enable peer-led recovery.
4. In cases where the researcher is not a permanent member of staff or volunteer of an organisation the idea of mixed loyalties might be better expressed as 'mixed responsibilities' to denote the temporal nature of the relationship. However, in organisational welfare settings, particular those which have become professionalised and embed a set code of conduct for volunteers to follow, the term mixed loyalties is appropriate in addressing the conflictual ethical decisions ethnographers must make performing their role as a volunteer. As noted above, 'mixed loyalties' is applicable to researcher subjectivity also so as to highlight the more durable and less fleeting ethical tensions that arise as part of the researcher-participant-field dynamic.
5. The notion of religious power more widely relates to the broad set of engagements with Foucault's examination of religion and its variants of discipline, confessional and pastoral power. These intersections are developed in considerably more depth than can be achieved here (see Carrette, 2000, 2013; Garmany, 2010).
6. For a discussion on masculinity and faith-based recovery spaces, see Hansen (2012).
7. In this way, practices of worship can be partly understood as a technology of the self that entails an acknowledgement of the relation of self to a Big Other which 'permit individuals to effect by their own means or with the help of others a certain number of operations on their own bodies and souls, thoughts, conduct, and way of being, so as to transform themselves in order to attain a certain state of happiness, purity, wisdom, perfection, or immortality' (Foucault, 1988, p. 18).
8. For a critical account on the discourse of 'recovery' and its co-constitution via moral, religious, psychological, criminal and medical registers, see White (2005).
9. These practical experiences also led to a wider belief that the criminalisation of drug use is short-sighted as it is counterproductive (Corva, 2008), often pushing people with complex psychological and medical problems further away from accessing the right support services, and perpetuates the abject figure of 'the addict' that gives legitimacy for the further retrenchment and moralisation of existing welfare support.
10. Academic criticism of the workings of religious power within ambivalent dynamics of care/control is a necessary but not entirely sufficient account of the complex ethical space of Hebron's therapeutic community. While my focus here has been the ethnographic challenges entailed in navigating the values and organisational practices of Hebron, faith-based therapeutic communities invite supplementary readings that better bring to the fore the affective and ethical geographies that evade easy categorisation under conventional grammars of analysis (May & Cloke, 2014). This includes the therapeutic and transformative potentialities among both staff and residents, the performativity of religious experience, and the sincere commitment of individuals to spend much of their lives living *with* and caring *for* marginal and highly stigmatised populations, and in doing so, take vows to dwell in semi-monastic community that seems counter-cultural to the pervasive values in society: for example, a vow of material poverty, shared finance, and chastity for non-married members. For some staff and residents, this might be understood in terms of 'protecting' their own recovery and that of others; but nevertheless this provides an important counterweight to any simple portrayal of Hebron and its organisational practices as a 'total institution' in Goffman's terms.

Acknowledgements

I would like to thank members of the Hebron community who participated in this research, for their helpful guidance and for welcoming me into their community. I also wish to thank Nadia von Benzon, Richard Gale, Paul Cloke, Mark Jayne, Richard Gorman, Geoff DeVerteuil, as well as Rob Wilton and the three anonymous reviewers for their insightful comments and suggestions. All oversights, however, remain my own.

Disclosure statement

No potential conflict of interest was reported by the author.

References

Baumbusch, J. L. (2011). Conducting critical ethnography in long-term residential care: Experiences of a novice researcher in the field. *Journal of Advanced Nursing, 67*, 184–192. doi:10.1111/j.1365-2648.2010.05413.x

Bell, L., & Nutt, L. (2002). Divided loyalties, divided expectations: Research ethics, professional and occupational responsibilities. In M. Mauthner, M. Birch, J. Jessop, & T. Miller (Eds.), *Ethics in qualitative research* (pp. 71–91). London: Sage.

Berridge, V. (1979). Morality and medical science: Concepts of narcotic addiction in Britain. *Annals of Science, 36*, 67–85. doi:10.1080/00033797900200131

Birdwell, J. (2013). *Faithful providers*. London: Demos.

Bourgois, P. (1998). The moral economies of homeless heroin addicts. *Substance Use and Misuse, 33*, 2323–2351.

Brandes, S. (2002). *Staying sober in Mexico city*. Austin: University of Texas Press.

Carlson, R., Singer, M., Stephens, R., & Sterk, C. (2009). Reflections on 40 years of ethnographic drug abuse research: Implications for the future. *Journal of Drug Issues, 39*, 57–70. doi:10.1177/002204260903900106

Carrette, J. (2000). *Foucault and religion*. London: Routledge.

Carrette, J. (2013). Foucault, religion, and pastoral power. In C. Falzon, T. O'Leary, & J. Sawicki (Eds.), *A companion to Foucault* (pp. 368–383). Chichester: Wiley-Blackwell.

Chenhall, R. (2008). What's in a rehab? Ethnographic evaluation research in indigenous Australian residential alcohol and drug rehabilitation centres. *Anthropology & Medicine, 15*, 105–116. doi:10.1080/13648470802122024

Chong, K. (2008). Coping with conflict, confronting resistance: Fieldwork emotions and identity management in a South Korean evangelical community. *Qualitative Sociology, 31*, 369–390. doi:10.1007/s11133-008-9114-0

Cloke, P., Cooke, P., Cursons, J., Milbourne, P., & Widdowfield, R. (2000). Ethics, reflexivity and research: Encounters with homeless people. *Ethics, Place and Environment, 3*, 133–154. doi:10.1080/713665889

Cloke, P., May, J., & Johnsen, S. (2010). *Swept-up lives: Re-envisioning the homeless city*. Chichester: Wiley-Blackwell.

Corva, D. (2008). Neoliberal globalization and the war on drugs: Transnationalizing illiberal governance in the Americas. *Political Geography, 27*, 176–193. doi:10.1016/j.polgeo.2007.07.008

Costley, C., Elliott, G., & Gibbs, P. (2010). *Doing work based research: Approaches to enquiry for insider-researchers* (pp. 48–62). London: Sage.

Coy, M. (2006). This morning I'm a researcher, this afternoon I'm an outreach worker: Ethical dilemmas in practitioner research. *International Journal of Social Research Methodology, 9*, 419–431. doi:10.1080/13645570601076785

Crowhurst, I. (2013). The fallacy of the instrumental gate? Contextualising the process of gaining access through gatekeepers. *International Journal of Social Research Methodology, 16*, 463–475. doi:10.1080/13645579.2013.823282

Davies, D. (2004, August 17). Charity's funds cut over religious links. *The Western Mail*. Retrieved from http://www.religionnewsblog.com/8382/charitys-funds-cut-over-religious-links
Dean, M. (1995). Governing the unemployed self in an active society. *Economy and Society, 24*, 559–583. doi:10.1080/03085149500000025
DeVerteuil, G. (2004). Systematic inquiry into barriers to researcher access: Evidence from a homeless shelter. *The Professional Geographer, 56*, 372–380. doi:10.1111/j.0033-0124.2004.05603006.x
DeVerteuil, G., & Wilton, R. (2009). Spaces of abeyance, care and survival: The addiction treatment system as a site of 'regulatory richness'. *Political Geography, 28*, 463–472. doi:10.1016/j.polgeo.2009.11.002
Disney, T. (2015). Complex spaces of orphan care – A Russian therapeutic children's community. *Children's Geographies, 13*, 30–43. doi:10.1080/14733285.2013.827874
Doyle, L. (1999). 'The big issue': Empowering homeless women through academic research? *Area, 31*, 239–246. doi:10.1111/j.1475-4762.1999.tb00088.x
Drake, D. H., Earle, R., & Sloan, J. (Eds.). (2015). *The Palgrave handbook of prison ethnography*. New York, NY: Palgrave Macmillan.
Duff, C. (2010). Enabling places and enabling resources: New directions for harm reduction research and practice. *Drug and Alcohol Review, 29*, 337–344. doi:10.1111/j.1465-3362.2010.00187.x
Evans, J. (2012). Supportive measures, enabling restraint: Governing homeless 'street drinkers' in Hamilton, Canada. *Social & Cultural Geography, 13*, 185–200. doi:10.1080/14649365.2012.655769
Evans, J., Semogas, D., Smallwood, J., & Lohfeld, L. (2015). 'This place gives me a reason to care': Understanding managed alcohol programs as enabling places. *Health & Place, 33*, 118–124. doi:10.1016/j.healthplace.2015.02.011
Foucault, M. (1988). Technologies of the self. In L. Martin, H. Gutman, & P. Hutton (Eds.), *Technologies of the self: A seminar with Michel Foucault* (pp. 16–49). Amherst: University of Massachusetts Press.
Foucault, M. (2001). *Hermeneutics of the subject: Lectures at the College de France, 1981–82*. F. Gros (Ed.) (G. Burchell, Trans.). Basingstoke: Palgrave Macmillan.
Fraser, S., & Valentine, K. (2008). *Substance and substitution: Methadone subjects in liberal societies*. New York, NY: Palgrave Macmillan.
Fuller, D. (1999). Part of the action, or 'going native'? Learning to cope with the 'politics of integration'. *Area, 31*, 221–227. doi:10.1111/j.1475-4762.1999.tb00086.x
Garmany, J. (2010). Religion and governmentality: Understanding governance in urban Brazil. *Geoforum, 41*, 908–918. doi:10.1016/j.geoforum.2010.06.005
Garthwaite, K. (2016). The perfect fit? Being both volunteer and ethnographer in a UK foodbank. *Journal of Organizational Ethnography, 5*, 60–71. doi:10.1108/JOE-01-2015-0009
Gill, N. (2016). *Nothing personal? Geographies of governing activism in the British asylum system*. Chichester: Wiley-Blackwell.
Goffman, A. (2015). *On the run*. Chicago, IL: University of Chicago Press.
Goffman, E. (1968). *Asylums*. Harmondsworth: Penguin.
Greener, J. (2015). Embedded neglect, entrenched abuse: Market failure and mistreatment in elderly residential care. *Social Policy Review, 27*, 131–149. doi:10.1332/policypress/9781447322771.003.0007
Guillemin, M., & Gillam, L. (2004). Ethics, reflexivity, and 'ethically important moments' in research. *Qualitative Inquiry, 10*, 261–280. doi:10.1177/1077800403262360
Han, J. H. J. (2010). Neither friends nor foes: Thoughts on ethnographic distance. *Geoforum, 41*, 11–14. doi:10.1016/j.geoforum.2009.09.007
Hansen, H. (2012). The 'new masculinity': Addiction treatment as a reconstruction of gender in Puerto Rican evangelist street ministries. *Social Science and Medicine, 74*, 1721–1728. doi:10.1016/j.socscimed.2011.06.048
Holloway, J., & Valins, O. (2002). Editorial: Placing religion and spirituality in geography. *Social & Cultural Geography, 3*, 5–9. doi:10.1080/14649360120114107
Hopkins, P. (2007). Positionalities and knowledge: Negotiating ethics in practice. *ACME: An International E-Journal for Critical Geographies*, 6, 386–394. Retrieved from http://ojs.unbc.ca/index.php/acme/article/view/787
Iveson, K. (2010). Some critical reflections on being critical: Reading for deviance, dominance or difference? *City, 14*, 434–441. doi:10.1080/13604813.2010.496213

Jayne, M., Holloway, S., & Valentine, G. (2008). Geographies of alcohol, drinking and drunkenness. *Progress in Human Geography, 32*, 247–263. doi:10.1177/0309132507087649

Jordon, M. (2014). *Convulsing bodies: Religion and resistance in Foucault*. Stanford, CA: Stanford University Press.

Lawhon, M., Herrick, C., & Daya, S. (2014). Researching sensitive topics in African cities: Reflections on alcohol research in Cape Town. *South African Geographical Journal, 96*, 15–30. doi:10.1080/03736245.2014.896279

Levine, H. (1978). The discovery of addiction. *Journal of Studies on Alcohol, 15*, 493–506. doi:10.15288/jsa.1978.39.143

Love, M., Wilton, R., & DeVerteuil, G. (2012). 'You have to make a new way of life': Women's drug treatment programmes as therapeutic landscapes in Canada. *Gender, Place & Culture, 19*, 382–396. doi:10.1080/0966369X.2011.609985

May, J., & Cloke, P. (2014). Modes of attentiveness: Reading for difference in geographies of homelessness. *Antipode, 46*, 894–920. doi:10.1111/anti.12043

Mjåland, K. (2015). The paradox of control: An ethnographic analysis of opiate maintenance treatment in a Norwegian prison. *International Journal of Drug Policy, 26*, 781–789. doi:10.1016/j.drugpo.2015.04.020

Mohammad, R. (2001). 'Insiders' and/or 'outsiders': Positionality, theory and praxis. In M. Limb & C. Dwyer (Eds.), *Qualitative methodologies for geographers: Issues and debates* (pp. 101–120). London: Arnold.

Moore, D. (2004). Governing street-based injecting drug users: A critique of heroin overdose prevention in Australia. *Social Science & Medicine, 59*, 1547–1557. doi:10.1016/j.socscimed.2004.01.029

Moran, D. (2015). *Carceral geography: Spaces and practices of incarceration*. Farnham: Ashgate.

Neff, J. A., Shorkey, C. T., & Windsor, L. C. (2006). Contrasting faith-based and traditional substance abuse treatment programs. *Journal of Substance Abuse Treatment, 30*, 49–61. doi:10.1016/j.jsat.2005.10.001

O'Neill, K. (2014). On liberation: Crack, christianity and captivity in post-war Guatemala city. *Social Text, 32*, 11–28. doi:10.1215/01642472-2703833

Parr, H. (2000). Interpreting the 'hidden social geographies' of mental health. *Health & Place, 6*, 225–237. doi:10.1016/S1353-8292(00)00025-3

Parr, H. (2003). Medical geography: Care and caring. *Progress in Human Geography, 27*, 212–221. doi:10.1191/0309132503ph423pr

Pollard, R. (2008, March 17). Lives at risk when vulnerable patients taken in by cult-like groups. *The Sydney Morning Herald*. Retrieved October 10, 2015, from http://www.smh.com.au/news/national/lives-at-risk-when-vulnerable-patients-taken-in-by-cultlike-groups/2008/03/16/1205602195125.html

Pullinger, J. (1980). *Chasing the dragon*. London: Hodder and Stoughton.

Sanchez, Z., & Nappo, S. A. (2008). Religious treatments for drug addiction: An exploratory study in Brazil. *Social Science & Medicine, 67*, 638–646. doi:10.1016/j.socscimed.2008.04.009

Sandberg, S., & Copes, H. (2013). Speaking with ethnographers: The challenges of researching drug dealers and offenders. *Journal of Drug Issues, 43*, 176–197. doi:10.1177/0022042612465275

Seddon, T. (2010). *A history of drugs*. London: Routledge.

Shaw, I., & Lunt, N. (2011). Navigating practitioner research. *British Journal of Social Work, 41*, 1548–1565. doi:10.1093/bjsw/bcr025

Skoll, G. (1992). *Walk the walk and talk the talk: An ethnography of a drug abuse treatment facility*. Philadelphia, PA: Temple University Press.

Thiem, C., & Robertson, M. (2010). Behind enemy lines: Reflections on the practice and production of oppositional research. *Geoforum, 41*, 5–6. doi:10.1016/j.geoforum.2009.11.001

Thomas, Y. F., Richardson, D., & Cheung, I. (Eds.). (2008). *Geography and drug addiction*. Dordrecht: Springer.

Tinney, J. (2008). Negotiating boundaries and roles: Challenges faced by the nursing home ethnographer. *Journal of Contemporary Ethnography, 37*, 202–225. doi:10.1177/0891241607312487

Valverde, M. (1998). *Diseases of the will*. Cambridge: Cambridge University Press.

White, W. (2005). Recovery. *Alcoholism Treatment Quarterly, 23*, 3–15. doi:10.1300/J020v23n01_02

Wilkerson, D. (1963). *The cross and the switchblade*. London: Marshall Pickering.

Williams, A. (2012). *Geographies of faith, welfare and substance abuse: From neoliberalism to postsecular ethics* (Unpublished doctoral thesis). Exeter: University of Exeter.

Williams, A. (2013). Practical theology and Christian responses to drug use. In P. Cloke, J. Beaumont, & A. Williams (Eds.), *Working faith: Faith-based organizations and urban social justice* (pp. 47–65). Milton Keynes: Paternoster Press.

Williams, A. (2016). Spiritual landscapes of Pentecostal worship, belief, and embodiment in a therapeutic community: New critical perspectives. *Emotion, Space and Society, 19*, 45–55. doi:10.1016/j.emospa.2015.12.001

Williams, A., Cloke, P., May, J., & Goodwin, M. (2016). Contested space: The contradictory political dynamics of food banking in the UK. *Environment and Planning A*, 1–26. doi:10.1177/0308518X16658292

Williams, M. (2016). Justice and care in the city: Uncovering everyday practices through research volunteering. *Area*, 1–8. doi:10.1111/area.12278.

Wilton, R., & DeVerteuil, G. (2006). Spaces of sobriety/sites of power: Examining social model alcohol recovery programs as therapeutic landscapes. *Social Science and Medicine, 63*, 649–661. doi:10.1016/j.socscimed.2006.01.022

Wilton, R., DeVerteuil, G., & Evans, J. (2014). 'No more of this macho bullshit': Drug treatment, place and the reworking of masculinity. *Transactions of the Institute of British Geographers, 39*, 291–303. doi:10.1111/tran.12023

Wilton, R., & Moreno, C. (2012). Critical geographies of drugs and alcohol. *Social & Cultural Geography, 13*, 99–108. doi:10.1080/14649365.2012.670505

OPEN ACCESS

Confessions of an inadequate researcher: space and supervision in research with learning disabled children

Nadia von Benzon

ABSTRACT

Location is often at the fore of decision-making regarding fieldwork and choice of methods. However, little research has directly discussed the importance of the choice of site in the production of research data, particularly concerning the way that different relationships will manifest between researcher and participant in different spaces. Site may be particularly important in research with (learning disabled) children, as research location is intertwined with the level of caregiving required from the researcher, and the sorts of surveillance the research engagement may be subject to. This paper draws on research with learning disabled 6–16-year olds that took place in homes, schools and the outdoors, in a variety of microgeographical locations from bedrooms to nature reserves. This paper reflects on the challenges, including the very 'worst' research moments, occurring in the different research environments. Whilst the research was carried out with learning disabled children and young people, the discussion has implications for research with non-disabled children and 'vulnerable' participants more broadly.

Confessions d'un chercheur inadéquat: espace et supervision dans la recherche auprès d'enfants ayant des difficultés d'apprentissage

RÉSUMÉ

Le lieu est souvent une priorité dans la prise de décision concernant le travail sur le terrain et le choix des méthodes. Pourtant, peu de recherches ont discuté directement l'importance du choix du lieu dans la production de données de recherche, en particulier en ce qui concerne la façon dont différents rapports se manifesteront entre le chercheur et le participant dans des espaces différents. Le lieu peut être particulièrement important dans la recherche auprès des enfants (avec des difficultés d'apprentissage) car le lieu de la recherche est étroitement lié au niveau de soins requis de la part du chercheur et aux types de surveillance dont l'engagement de la recherche peut faire l'objet. Cet article suit Horton (2008. A 'sense of failure'? Everydayness and research ethics. *Children's Geographies*, 6, 4, 363-383.) parmi d'autres dans le domaine des géographies des enfants et des géographies sociales, qui se penchent sur les défis, y compris les « pires » moments de la recherche, se produisant dans différents environnements de la recherche. Bien que la recherche soit effectuée auprès d'enfants ayant des difficultés d'apprentissage et de jeunes gens, la discussion a des implications pour la recherche auprès d'enfants non handicapés et de participants « vulnérables » plus généralement.

Confesiones de un investigador inadecuado: espacio y supervisión en la investigación con niños con dificultades de aprendizaje

RESUMEN
La ubicación es a menudo uno de los factores principales en la toma de decisiones en relación con el trabajo de campo y la elección de métodos. Sin embargo, pocos estudios han examinado directamente la importancia de la elección del sitio en la producción de datos de investigación, en particular sobre la forma en que las diferentes relaciones se manifestarán entre el investigador y participante en diferentes espacios. El sitio puede ser particularmente importante en la investigación con niños (con dificultades de aprendizaje), ya que la ubicación de la investigación se entrelaza con el nivel de prestación de cuidados que se requiere por parte del investigador, y el tipo de vigilancia al que puede estar sujeto el compromiso de la investigación. Este documento se basa en la investigación con niños con dificultades de aprendizaje de entre 6 y 16 años que se llevó a cabo en hogares, escuelas y espacios abiertos, en una variedad de lugares micro-geográficos, desde habitaciones hasta reservas naturales. El documento sigue a Horton (2008. A 'sense of failure'? Everydayness and research ethics. *Children's Geographies*, 6, 4, 363-383.), entre otros investigadores en la geografía de los niños y geografías sociales, al reflexionar sobre los desafíos, incluyendo los 'peores' momentos de investigación, que se producen en diferentes entornos de investigación. Mientras que la investigación se llevó a cabo con niños y jóvenes con dificultades de aprendizaje, la discusión tiene implicaciones para la investigación con niños sin dificultades y los participantes 'vulnerables' en términos más generales.

Introduction

This paper examines the role of location in a recent study involving learning disabled children and young people as research participants, exploring perception and experiences of green spaces. Whilst research with this participant group throws up a wide variety of methodological, ethical and practical challenges, this paper focuses particularly on the spaces in which research was conducted, the negotiations that occurred over control of these spaces and although rarely explicit, over control of the participants in these spaces. Sites of research will be considered in terms of the way in which they define, or at least influence, the relationship between the researcher and the participant, particularly in the light of mediation, or involvement, from third-party gatekeepers. This approach builds on the burgeoning interest in research ethics, and their practical and methodological implications, evident throughout human geography publications of the last 20 years. These discussions have certainly been at the fore of children's geography where the ethical implications of research are increasingly given as much concern as the empirical or theoretical outcomes (see e.g. Horton, 2001).

Whilst interest is growing, space in which qualitative research occurs, and the ethical and methodological implications over the characteristics of this environment, has not been discussed as rigorously as other aspects of conducting research (Jones, 2008). This paper considers the implications of the variety of spaces that were offered as potential research locations in participants' homes, in school and when conducting research outdoors or in alternative venues. Each sort of space offered different opportunities and challenges, particularly in terms of avoiding both isolation and interference, requiring a different set of behaviours and practices from the researcher and resulting in a unique research relationship. This paper intends to present a frank discussion of the ethical and methodological implications of the choices and compromises over research space made by the author when researching with learning disabled children and young people, aged 6–16. This discussion is predicated upon three vignettes illustrating some of the 'worst' moments of my research experiences. These are moments in which my actions, or failure to act, led to harm for the participants – moments I am ashamed of, or at least confused about.

Focusing on negative research experiences is not a novel way to explore research ethics and methods. For example, Horton (2008) draws on Rose (1997) in considering moments of failure. Horton (2008) argues that this approach highlights the way in which the messiness of everyday life, acknowledged by recent developments in social and cultural geography, is reproduced in the research process. Moreover, he argues that acknowledging the messiness and unpredictability of the real world presents significant challenges to a traditional, static, approach to research ethics. This paper reflects on my own experiences of failure in order to produce a constructive response to the inadequacy I felt in those research moments; as Pyer (2008, p. 216) states: 'it is not good enough to rationalise the unintended simply by saying it was unfortunate – ethical research depends on more than that'. In choosing failures as a topic of deliberation, this paper intends to highlight the power relations manifest in the spaces of research that might not have been so starkly illustrated, had the young people encountered a positive experience within the same networks of governance. This is to say that when research processes go to plan, embedded relationships of power may be invisible. When expectations are conflicted, or processes disrupted, previously unnoticed, or unspoken, hierarchies may be illuminated. As such, reflection on these embedded relationships may be sufficient to turn these 'unfortunate' events into useful learning experiences.

The research that is the foundation of this discussion, occurred as part of a social geography investigation, but was situated in the context of an understanding of the social model of disability. The social model is a key, although not uncriticised, approach within Disability Studies and allied disciplines, with its roots in disability activism and disability rights dialogue. The social model states that people are disabled, not as a logical outcome of impairment, but through society's inability to meet their individual needs (see e.g. Oliver, 2004). For this reason, the UK Disability Studies movement favours the use of the term 'disabled person', as it emphasises the fact that society has done the disabling, rather than 'person with a disability', which locates the 'fault' with the individual. In addition to the linguistic ramifications of this approach, my research with learning disabled children has also sought to reproduce some of the political discourse of the social model. This has involved attempts to maximise the agency of the child participants, adapting methods and approaches to draw on their strengths in communication and promoting an agenda of choice, with an ideal that the participants will become the key decision-makers in the research process. This approach is also very much in line with the participatory approaches at the fore of research in children's geography and across social geography (Abebe & Bessell, 2014; Alderson, 1995; Cope, 2008, amongst many others).

Place in research with children

Decisions about the location of research activities are often central to discussions in the research planning stage of the fieldwork process. Initial discussions concerning location tend to centre around the practicalities of accessing the intended participants in the chosen space – normally relating to the number of participants needed for the study, the duration and frequency of contact and the potential to engage the participants in the sorts of activities desired by the researcher. However, these early discussions may also consider an assessment of the risks involved, and the 'ethical' implications of conducting research in a particular site (Anderson & Jones, 2009; Bushin, 2007; Elwood & Martin, 2000; Valentine, 1999, amongst many others).

I emphasise 'ethical', in agreement with Halse and Honey (2007) and Allen (2009) regarding the gulf in understanding that might exist between 'ethics' as comprehended by an appointed institutional committee, and the 'ethical' imperative of a social science researcher aspiring to a participatory or emancipatory approach. By this I mean that a traditional approach to ethics, and the one that still appears to be the working paradigm for many ethics committees, is that ethical research should minimise harm (Halse & Honey, 2007). This is in terms of both physical and psychological harm, to both participants and researchers. However, a participatory approach to ethics suggests that our duty as researchers goes beyond preventing harm, to maximising agency amongst our participants (e.g. Cope, 2008) and/or maximising benefit to relevant communities (e.g. Mills, 2013). Unfortunately, the two approaches to ethics, an approach that prevents harm, and an approach that maximises benefit, may be, in children's geographies at least, at odds with one another (Abebe & Bessell, 2014; Hammersley, 2009). The clearest example of this is in terms of consent to participate, where researchers are often duty bound by ethics committees to obtain parental consent, perceived as a way of minimising harm. Conversely, children's geographers might argue that this simultaneously removes children's agency in the process, and therefore contradicts participatory ethics (see Masson, 2000). The widespread institutional reticence to engage with concerns over children's agency in research leads decisions over research venue to be viewed as a predominantly practical consideration, relating to 'accessibility' and 'safety', in their various forms, with little acknowledgement of any ontological implications (see also Dyer & Demeritt, 2009).

This paper intends to explore both the 'ethical' (relating to participant power in the research process) and the ontological significance of conducting research with children in a place. That is to say, the importance of recognising that research activities occur not only in a room, at a table, on some grass, but within a building or on land that has significance as being bounded by institutional or familial practices and norms. Elwood and Martin (2000, p. 649) identify the 'microgeography' of the interview site, arguing that decisions on the location of research can illustrate the social geographies of a place. In this paper, research sites will be reconstructed as research places with imbibed ways of being and ways of behaving that may have significant influence on the interactions between researcher and participant, and the sorts of knowledge that are produced through this relationship. Focusing on learning disabled children as research participants may provide a particularly acute example of a working relationship that is a product of, or at least heavily influenced by, the place in which it occurs. This is due to the heightened social perception of the vulnerability and incapacity of learning disabled children, and the high levels of protection duly afforded to them (for a classic discussion of this issue, see Wolpert, 1980; and for more detail concerning my own participants, von Benzon, 2011). However, foregrounding discussion about the dynamics of researcher–participant interactions and the importance of the environment in which they occur have significant implications for research with all 'vulnerable' participants, and even those typically felt to have relative agency in the research process.

The growth in interest in methods and practice in research with children is part of a wider academic focus, since the 1990s, on the ethics of undertaking research with a population understood to be marginalised (Alderson, 1995; Morrow & Richards, 1996; Rose, 1997; Valentine, 1999; Winchester, 1996, amongst many others). The recognition of the unequal power relations present in social interaction, that manifest in a research relationship between an adult researcher and a child participant, is by no means novel. What is new, or at least, increasingly present, is the acceleration of discourse concerning the ethics of research with children relating to all facets of the research process. Particularly, this is in relation to the discussion of strategies to overcome ethical, methodological and practical challenges (e.g. Bushin, 2007), and honesty and reflexivity in the voice of the author (e.g. Horton, 2008). Despite the academic interest in the practicalities of carrying out research with children and young people, it is often difficult to tease out the importance of site in research with children. Even papers that focus in some detail on ethical and methodological aspects of research rarely devote space to discussion of the ontological importance of the research site. There are notable exceptions, of course (such as Elwood & Martin, 2000; Jones, 2008), which will be explored. The following section considers in turn literature relating to research in schools, in homes and in 'alternative' sites.

Research in school

Schools provide the location for much research with children (e.g. Barker, Alldred, Watts, & Dodman, 2010; Davis, Watson, & Cunningham-Burley, 2000; Goodfellow, 2012; Holt, 2003; Jones, 2008, amongst many others). Once the researcher is granted entry, schools offer potential access to large numbers of children for long periods of time. Other than the practical advantages, there may be clear epistemological arguments for undertaking research within a school. School is children's work; children's lives tend to centre around school, with their friendships often revolving around their school-based social circles (Morris-Roberts, 2004). For many children, schools are the key space in which their intellect, sociability and physique are developed, and critiqued (Hemming, 2011; Holt, 2004); therefore, research that occurs in school might be argued to be better placed to catch sight of these moments in children's lives. It might also be argued that school-based research has greater potential socio-political impact, as school's policy and practices are heavily influenced, determined even, by outside agencies who may in turn be influenced by research outcomes in a way that family behaviours may not (e.g. Evans, Evans, & Rich, 2003).

The school as a research site might have particular significance for learning disabled children, whether attending a segregated specialist school, or a mainstream school. Either mode of education will result in the young person experiencing difference, whether through segregation from their peers in the local community, through part- or full-time use of a specialist unit or as someone who receives additional support within a mainstream classroom (Goodfellow, 2012; Holt, 2003; McMaugh, 2011; Valentine & Skelton, 2003). Not only are young disabled people likely to be positioned as 'other' as a result of their immediate physical or intellectual differences, and the specific educational spaces they access, they are also likely to be differentiated by limited independence within school (Worth, 2013). Disabled young people are likely to be more reliant on support, from both technological aids and other people, than their non-disabled peers (Costley, 2000).

Many researchers would argue that a school is not simply *a* school. By which I mean that there is no single school environment as schools contain a vast array of microgeographies of different spaces demanding different sorts of behaviours and social relations (Elwood & Martin, 2000; Holt, 2004). For example, the lived experience of the playground will contrast to the classroom, the dining hall, the gym, the science lab or even the store cupboard (Jones, 2008; Pike, 2010), due to differences in behavioural norms in these different spaces. The specific location of research within a certain microenvironment has been shown to have influence on the sorts of contributions that children will offer (Anderson & Jones, 2009). For example, Jones (2008) found that participants were happy to open up to her and share the personal details of their life in the intimate space of the store cupboard, in a manner in which they did not appear confident doing in the classroom. Holt (2004) discusses microgeographes in a school in terms of spheres of power, with different teachers implementing school policies in different ways, allowing for different behaviours and relationships in different classrooms.

Whilst the microgeography of the location is important, the temporality of the location is also a key characteristic of the research engagement. Valentine (1999) suggests that in the school context, researchers can benefit from institutional norms where children are expected to comply with adult demands (see Horton, 2008). For example, Robson and Ansell (2000, p. 178) describe conducting research 'during a lesson period when the usual teacher was absent ... organised in a manner which conformed to students' expectations of a lesson format'. Children's geographers are typically striving to promote the agency of participants, meanwhile undertaking research as part of a child's mandatory lessons, all but removes this agency (Denscombe & Aubrook, 1992). In a research context ultimately focused on producing interesting research data, it takes a strong will for a children's researcher not to attempt to make use of this environment of compliance to ensure research participation and data generation.

Research in the home

Whilst Anderson and Jones (2009) illustrate the variation in environment within schools, homes can also offer a wide range of spaces for research. Differences in the microenvironments of the home may be more limited than those in a large institution. However, spaces in the home afford variations in their levels of privacy, and in the sorts of behaviours expected in them. For example, Mauthner (1997) describes the way in which she was led by a child, away from the lounge which was 'public' and characterised by comings and goings, to their own bedroom, so that sensitive issues could be discussed privately. Similarly, Nilsen and Rogers (2005) discuss the demarcation of private and public spaces within the home. Parents didn't show the researchers their bedrooms, despite freely offering tours of the rest of the house. Meanwhile, children used the research visits as a pretext to transgress boundaries and enter parents' or siblings' bedrooms. These microgeographies of the home have been less extensively explored than the microgeographies of the school, reflective perhaps of a broader discrepancy between the quantity of research conducted in schools and in homes (Bushin, 2007; Nilsen & Rogers, 2005).

Most research in households focuses on research in the child's own home, within their family (e.g. Punch, 2007; Starkweather, 2012; Tyrrell & Harmer, 2015). Whilst the home might be normatively constructed as a space of comfort, security and familiarity, a number of children's geographers have conducted research in spaces that do not conform to this ideal. Some children's geography research, focused in the global South, has involved child residential domestic workers being interviewed in their employers' homes (Klocker, 2012). Meanwhile, Dorrer, McIntosh, Punch, and Edmond (2010) conduct research in residential children's homes exploring food practices. On the other hand, research apparently undertaken in a space that does not conform to the researcher's ideas of home may provide that sense of safety or belonging to the participants. For example, Ursin (2011) found the young homeless Brazilians in his research use the term 'home' to refer to their sleeping space in the street. There may be further methodological and ethical questions arising when young people are interviewed in other's homes (Tucker & Matthews, 2001), and in residential care settings (Winter, 2012).

The multiplicity of spaces that might be considered as 'home', and the 'like home' relationships with space that may exist for participants make defining, and even categorising, research in the home, a complex discursive challenge (see Nilsen & Rogers, 2005). Despite this challenge, research in the home has been credited with offering children more agency than research in schools, as a site in which children are more likely to be involved in decision-making on a day-to-day basis (Bushin, 2007). As such, the clear conflict between the expectation that children follow adult instructions and the researcher's desire to promote children's agency that occurs within the school may be less apparent at home. However, the extent to which this is true is likely to vary between households and families, with more variation in practice likely between different households than between different schools (see Hood, Kelley, & Mayall, 1996). One measure of children's agency in their home might be the extent to which children are involved in decision-making regarding research participation in the research project itself (Lewis, 2009). For example, Bushin (2007) found that despite being told a child participant had agreed to take part, the child was in fact unaware of the planned research visit, an experienced mirror in my own research visits.

There is reason to believe that lack of agency within the home may be a particular issue for disabled children. The majority of disabled children in Britain now live in the community, with their biological family (Contact A Family, n.d.). However, statistics show that disabled children's households score highly on a number of indicators for deprivation. For example, disabled children are significantly more likely to grow up in poverty than non-disabled children (Shahtahmasebi, Berridge, Lancaster, & Emerson, 2011), with 31% of UK families with disabled children without money to buy sufficient food (Contact A Family, n.d.). As such, families with disabled children can broadly be considered as particularly economically disadvantaged. This socio-economic disadvantage may combine with an overall lack of agency in the institutional service provision that surrounds their disabled child, creating a lived experience of powerlessness amongst families of disabled children.

Where families of disabled children are not more broadly disadvantaged, disabled children within these families may still experience restrictions in agency relative to their non-disabled peers. For example, older disabled children and teenagers may experience reduced independent mobility in order for parents to minimise perceived risk (Jahonda, Wilson, & Stalker, 2010). However, lack of agency can also be brought about by social stigma and marginalisation that leads disabled children to choose to spend far more time at home, and alone, than their non-disabled siblings (Buttimer & Tierney, 2005; Cooney, Jahoda, Gumley, & Knott, 2006; Solish, Perry, & Minnes, 2010). As such, the home may take on a different significance for disabled children, being at once a refuge, and a site of isolation. These factors have implications for researchers intending to enter the homes of disabled children and their families.

Alternative spaces

'Alternative' spaces, defined here as: the spaces that do not exist within the bounds of the school or the home, commonly appear in the literature in two ways: mobile interviews and interviews that are organised directly with young people, outside of institutions. The first approach includes methods such as 'interviews-in-situ' (Anderson & Jones, 2009), place-based interviewing (Holton & Riley, 2014), 'the walk-around' (Griffin, Lahman, & Opitz, 2014), 'emplaced' (Booth, 2015) and 'Go-along' (Bell, Phoenix, Lovell, & Wheeler, 2015). These approaches to data collection are commonly valued as means of 'elighten'ing and 'enliven'ing research (Holton & Riley, 2014), being argued to facilitate depth of data production, as the embodied experience of the place encourages the elicitation of memories and ideas.

Another approach to defining 'alternative' spaces in research with children is to consider them as those places that are not accessed through negotiations with schools or families. Children's research addressing non-institutional spaces, especially 'the street', is common in an international, particularly a majority world, context (Ursin, 2011; van Blerk, 2005) but not the norm in a North American or European context. This is due to a variety of reasons, most significantly perhaps due to the institutionalisation of childhood throughout the minority world, and the suspicion that falls on unregulated adult–child interactions (Holloway & Pimlott-Wilson, 2014; Malone, 2007). Most researchers, and research institutions, would consider an adult researcher too vulnerable to misinterpretation or accusation if they were to directly approach children. A good illustration of this reticence is provided by Matthews, Limb, and Taylor (2000); to explore young people's use of 'the street', they sought written permission from parents, and qualified as outreach youth workers, before young people were approached. More generally, research addressing young people's use of public space, or independent, uninstutionalised movements, tends to be embedded in a research relationship based within some sort of institution (e.g. Collins & Kearns, 2001; Robinson, 2009).

There is a particular absence of disabled young people in research in 'alternative spaces'. This is both in terms of mobile research and research that occurs beyond institutional boundaries. The primary exceptions to this are work addressing mobility with mobility-impaired young people such as wheelchair users (Pyer and Tucker, 2014) and visually impaired teenagers (Worth, 2013). One significant barrier to mobile research with disabled young people may be ethics and risk assessment considering this too risky an endeavour. This was my own experience when planning to conduct research with learning disabled young people, as the ethics committee insisted on a second adult present in order to manage unruly or unpredictable behaviour form the young people. Similarly, gate keepers may consider researchers unqualified or inexperienced to adequately safeguard participants. The lack of disabled young people in research in 'alternative' spaces mirrors a broad lack of disabled young people engaging in social activities outside the home and school (Solish et al., 2010). As such, researchers intending to undertake research with disabled children may find it difficult to recruit participants, or may come across difficulties in entering these unbounded spaces with research participants.

Fieldwork

The remainder of the discussion in this paper is based on research experience gleaned primarily from two projects involving fieldwork with learning disabled young people aged 6–16. The first of these projects was undertaken in Oxfordshire, involving 11 learning disabled young people, aged 6–16, and 3 non-disabled siblings. The project was based in the children's homes and explored their usage and perceptions of green space. Participants took part in weekly activities, typically lasting from about 30–90 min, for approximately five weeks. During each session, participants were offered a choice from a range of activities devised by the researcher to facilitate discussion around experiences and perceptions of outdoor green spaces. Initial activities included things like drawing, collaging, reading stories, games and keeping diaries. Activities were designed to develop a trusting relationship between the researcher, participant and their families, and so included increasingly mobile activities like playing in the garden and tours of the neighbourhood. Most participants took part in a final activity consisting of a trip to a nearby 'natural environment' venue, such as a bird of prey centre, an indoor rainforest or a wildlife park.

Research with the participants was organised by contacting the families directly. Following some years working for a charity providing recreational activities for learning disabled young people, I was able to advertise the research to users. Families wishing to participate were able to contact me directly and an initial informal discussion was arranged. In this manner, parents, and in the vast majority of cases, mothers, became the 'gatekeepers' to accessing the participants. Despite requesting parents discuss the project with their children prior to making contact, in most cases, parents had independently made the decision to involve their child in the project (cf. Bushin, 2007). Throughout the research process, parents continued to control the practicalities (cf. Nilsen & Rogers, 2005). It was with parents that I negotiated the timings of sessions, and also the sites in which the activities would occur. Research occurred in different spaces in different houses, often at the table in the dining room, or, more informally, on the floor in the sitting room. In one household, I was invited to use the kitchen table, and in another, we were sent to the child's bedroom.

The second project involved research engagement in a Special Educational Needs (SEN) secondary school in Greater Manchester for a few days each week, over the course of a year. Participants were approximately 100 young people aged 11–16 with mild–moderate learning disability. During the fieldwork period, I conducted ethnographic research, joining in with school activities on and off site, and devising some of my own outdoor activities for participants. I undertook active interviews with many of the pupils, making video diaries, running group games and class discussions and doing project work with older participants focusing on the development of an onsite nature area.

My presence as a researcher in school was negotiated initially with the deputy head teacher. In the first instance, I was invited into school to be interviewed, as a response to one of many 'cold calling' letters I had sent to SEN schools across Greater Manchester. As a researcher, I was then placed under the jurisdiction of a departmental head. I was quickly given the freedom to make contact with any staff in the school I wished to work with, and to impinge on classes to run research activities. About three months into fieldwork, I was invited to attend the week-long residential camp for 11–14-year olds run by the school each year. The staff proposed 'nature' as the theme of the camp, and I was welcomed to plan activities that would support my own research. Throughout the project, research occurred in a variety of spaces including the classroom, a conference room, the onsite nature area, various off site spaces and even our own classroom store cupboard (cf. Jones, 2008)!

The two projects provided experience of conducting research with learning disabled children and young people in a variety of locations in and around the school and home, and in 'alternative' spaces. Research was undertaken in the presence of teachers and teaching assistants, parents, siblings and, in a not-inconsequential minority of cases, alone with the participant, or group of participants. The remainder of the paper seeks to explore the spaces of these research relationships, positioning them in relation to the institutional and familial contexts in which they occurred. The discussion focuses on reflection on three individual and unconnected moments in research and their spatial contexts, illustrated by extracts from research data. These are not three vignettes chosen at random; rather, these stories are illustrations of some of my worst experiences as a researcher. I have chosen to use unpleasant research experiences, following children's geographers such as Horton (2008) and Pyer (2008), primarily for two reasons: firstly, as a means of highlighting the serious consequences of failure to address, or at least acknowledge, the significant imbalances of power that happen in research within institutions and family homes, and secondly, as a way of finding some 'value' in the mistakes I made. Throughout these reflections, children's identities are changed, but the facts are delivered as accurately as possible.

The day James accused me

James, a young teenager at the time, had been a particularly enthusiastic participant. I had been visiting him at home, once a week, and we'd been completing research tasks in the sitting room and dining room of his house. We'd had five sessions at home, and in his neighbourhood, before I took him out on a day trip. James had seemed to enjoy talking to me and completing activities like drawing pictures and cutting and sticking. Throughout, James was polite and amiable, keen to please and happy to turn his attention to tasks that I suggested, or choosing from a selection which he'd like to do. At times I was slightly concerned that James appeared to be a little over-friendly, telling me he thought I was very pretty, and drawing pictures in which I featured as his 'best friend'. However, I wasn't particularly worried and lightly brushed off these comments. We'd mostly been given space to work one-to-one at home, in the dining room or on the living room floor. James has epilepsy, which, at the time of the research, was largely kept under control by medication. However, his mum was used to providing a high level of care for him, and was concerned about him being too far away from her.

Due to James' mum's concerns, I had discussions with her regarding the choice of venue for our end-of-research green space visit, rather than choosing with the participant themselves, which had been the case for the other families. James' mum had dismissed many of the options I'd presented as being too far to travel, or being locations that risked triggering a fit. She recommended we visit a local nature reserve with a reservoir, woodland walks and a large play area. She was insistent that James loved it there, and would be delighted to go with me. James was aware that a friend of his, Tom, was also participating in the research project and asked if Tom could accompany us. Both sets of parents agreed. Also present on the trip was our 'chaperone', Lucy. Lucy was a police-checked student, known to me as a volunteer of a charity that I had previously worked for.

The drive to the nature reserve was uneventful. I collected Lucy, James and Tom in the car, and the boys seemed excited about our half-day out. It was the school holidays, so they were animated, and keen to stretch their legs. However, when we arrived at the reserve, the atmosphere took a turn for the worse. Both boys voiced their disappointment when we arrived, claiming that they 'came here all the time' that it was 'boring' and that there was 'nothing to do'. Lucy and I tried to jolly them along, setting out for a walk around the lake. The boys became increasingly excitable and began to fight, first verbally, and then hitting each other, throwing stones and sticks at each other, and at the ducks, and ignoring our pleas to 'calm down'. Eventually Lucy and I decided we were probably best off calling it a day and heading home.

At this suggestion the boys lost their tempers, arguing that they wanted to visit the shop and the playground, and running off. Lucy and I were at a loss, having no choice but to follow them. We spent a further hour or so cajoling, begging and threatening the boys to get them to return to the car, but not before they'd bought things in the shop with their pocket money, and had a long play in the playground. The return car journey was also fraught. The boys fought in the back of the car, threw food out of the car window, and kicked the back of our seats. At one point I pulled on to the hard shoulder of the dual carriageway to wait for them to calm down, as it seemed too unsafe to drive.

I dropped Tom off home with an exchange of pleasantries with him and his mum, and only a hint that his behaviour had been difficult to manage. When I dropped James off, I spent longer chatting to his dad, who was friendly and asked lots of questions about our afternoon out. Again, I didn't detail the problems of the afternoon but suggested that James had been fairly 'excitable', and that I didn't think the boys had liked the choice of venue.

As I turned to leave, James exclaimed, 'and Nadia hit me'. I was shocked at the accusation, and desperately worried that I would get in terrible trouble, despite never having laid a hand on either of the boys. James' dad looked surprised and concerned and told James that he didn't think that was true, and that James should think very carefully before making horrible accusations like that. He apologised to me for James' behaviour and I left, somewhat relieved, but still shaken. I spent the next few hours feeling rather sick, worried about the conversation that would take place in their house when James' mum got home.

Early that evening I received a call from James' mum. She was very apologetic, both about James' accusation, and his behaviour during the trip, which James had apparently told her about in detail. She said it wasn't the first time James had accused someone of hitting him when he felt embarrassed about his behaviour, as a way of deflecting blame. Despite my protestations, she was insistent that James was in a lot of trouble at home as a result of his behaviour that day, and that he would apologise further himself. A few days later I received a hand-made card from James thanking me for doing fun activities with him, and apologising for behaving badly on our trip out.

This episode in the research project throws up a variety of mistakes that I made as a researcher, highlighting the importance of site in research, particularly in terms of the way that location can change the research relationship. In the first instance, the visit represented a loss of control over the research process for James. Whilst he had been in control over decisions about activities at home, unlike other participants, James did not participate in decision-making over the venue for our trip out. He then reacted badly to realising we were in a place he considered 'boring'. The dynamics of our research relationship was also changed by the introduction of new people. This too was to some extent a function of the site, with the ethics committee requiring I have an additional adult to supervise. I had worried that two adults would be rather uncomfortable in a research visit with one child, so arranging to take two participants at once had seemed like a logical step. Unfortunately, as the dynamic changed from one-to-one research to a visit with two adults and two children, James seemed to see this as a move from a friendly working relationship to one in which our roles as 'adult' and 'child' were highly differentiated. This was compounded when James began to behave in a dangerous manner, and I felt compelled to intervene as a responsible adult.

This is not an entirely uncommon experience of children's geographers (see e.g. Punch, 2007; Morris-Roberts, 2001) who struggle in maintaining a 'least adult role' (Mandell, 1988) whilst managing difficult behaviour. Perhaps James' erratic and aggressive behaviour as a result of loss of control might have been pre-empted by more explicit discussion before the visit, about the way in which our research relationship might change in the new space. Similarly, we might have spoken in more detail about the activities that would happen on the visit, and James could have decided for himself the order in which activities would occur. I also think it was a mistake to have been guided so directly by James' mother over choice of venue. Perhaps a more comfortable decision-making process would have been to allow James' mother to choose two or three venues that she was happy with, and allowed James to have the final choice over site. In this way, he might have felt that he was retaining some control over the activity.

Ryan was forced to take part

As part of the school-based research project, I had arranged to do some small group work with 12–13 year old participants. I had suggested working with a group of four students over a period of four weeks, doing different activities. The science coordinator was keen on this idea, seeing it as a chance to develop social skills amongst the students who struggled most to engage in class and with each other. Wary that I was taking participants out of their regular class, I agreed to work initially with students chosen by the teacher. Four students were invited, whilst I was present in the classroom, and they all readily agreed. Other students were eager to have a turn as well, apparently seeing our research session as 'fun' and a chance to escape the normal lesson. The first session went well, and all four participants told me they were keen to do the same the following week.

However, the next week, three of the participants approached me, before class had begun, keen to get going. The fourth, Ryan, asked if he could miss the session this week and remain in class instead. I assured him that was no problem at all. The teacher arrived and I checked with her it was ok to leave, as per the previous week, with the participants. She noticed that Ryan was missing from the group, and asked him to join the research group. He and I both explained that he didn't feel like joining in this week, and that I had agreed it was absolutely fine for him to stay in class.

To my surprise, the teacher was insistent that Ryan join us. She said that he'd made a commitment to take part in the research, and that it wasn't fair on me, or the others, for him to pull out at this point. She insisted that he needed to learn to be responsible, and do things that he'd said he would. Ryan reluctantly left the room with us. In the school conference room in which research activities occurred, I apologised to Ryan and said that he was very welcome to sit to one side. He was keen to use the computers in the room, but I didn't risk letting him, as we didn't have permission. After a while, Ryan began to choose to join in with the research activities, as there really wasn't anything else for him to do in the room. Eventually he asked if he could go back to the classroom. I said that was fine, and he could tell his teacher that he'd finished the research activity early, and that I'd sent him back.

At the end of class, I returned with the other three students and found Ryan sitting separately from the rest of the class, at the back of the room. His teacher asked me if he had been naughty, and said that she was planning on keeping him in during lunch time. I assured her that he hadn't been behaving badly, and told her he'd just finished the activity early. She did let Ryan go off to lunch, but not before reiterating to him how important it was for him to do as he'd agreed. The next week Ryan left with us enthusiastically, and appeared to enjoy the session. Once the run of four sessions was complete, I chose not to repeat this method with other participants.

As with James, the problems for Ryan arose around lack of agency. Here, rather than the participants' mother controlling the situation, both the participant and the researcher became powerless in the face of the class teacher, who had authority in the institutional context. Ultimate control over the participants, as pupils in school, was expected by both the class teacher and the young people themselves, who are used to, and accepting of, a very high level of control over their bodies and behaviours within the classroom (Harden, 2012). However, whilst an adult, a professional and a peer, the researcher is also subject to the authority of the classroom teacher, who has the power to terminate the research project, in much the same way as a parent might ask you to leave their home (Davis et al., 2000). As such, a researcher within a school walks a difficult line between an ethical duty to support and respect children as independent social agents (Morris-Roberts, 2001), and satisfy class teachers that they are working within the ethos of the institution. The potentially uncomfortable power dynamic that exists between teacher and researcher is not one that is rigorously discussed in the literature.

Dan got sent home from camp

The residential science camp took place in May, about six months after I'd started working in school. Planning for the trip had begun a few months earlier, and I was approached in around March, by the camp co-ordinator, to participate. I was delighted, seeing it as a fantastic opportunity to produce some interesting research data with participants *in situ*. This enthusiasm increased further when the staff suggested in our initial camp prep meeting, that the theme of the trip should be 'nature' to fit with my research. I was to be allowed considerable freedom to organise research activities in addition to the variety of trips to 'nature-based' locations I'd be able to observe. The catch, that I was informed of a few weeks in to the planning, was that the deputy head had said I was only allowed to go if I officially took on a staff role during the trip, taking equal responsibility for the welfare of the participants.

The particular incident that precipitated Dan's removal from camp occurred when the students were playing in the outdoor 'pool' at our 'campsite'. We were staying in scout huts not far from Greater Manchester, and the site had an outdoor pool of sorts, more closely resembling a rectangular, concrete, pond. Dan was splashing about in the waist-deep water, and turned to me, smiling, and did an impression of someone being dragged under water by a shark. Standing on the bank, I laughed, and reached for my video camera. 'That was funny, do it again Dan, so I can get it on video' I called. Just as he began, two teachers saw him and began shouting at him to stop. I'm not sure whether he was too caught up in the moment, or whether he didn't hear them as he was splashing and shouting, but he continued, ignoring them. As soon as he had quietened down the teachers ordered him out of the pool, shouting that his behaviour was incredibly dangerous. I apologised, saying that I had asked him to do it. I was ignored, on the grounds that he had a long list of misdemeanours to his name. That evening Dan was sent home, the senior staff on the camp deciding he was a danger to himself and to others.

To some extent, the embedded power hierarchies of the classroom were blurred a little on camp. This was reflected through a change in titles, with teachers referred to by animal names, rather than their titles and surnames. Pupils wore their own clothes, rather than school uniforms throughout the week, and plenty of treats like hot chocolate, camp fires and ice creams added to creating an atmosphere that was different to that within school. Nevertheless, the desire to manage bodies that was embedded within the institutional boundaries of the school, and expected within the enclosed spaces of classrooms and hallways, spilled over into the less bounded and somewhat liminal spaces of camp. There had been no immediate danger to anyone from Dan's behaviour, he was well away from other students, and standing in water that wasn't more than waste deep, and yet Dan was excluded on the basis of failure to conform to the sort of normative bodily practices expected in the school environment, whilst in the outdoors (cf. von Benzon, 2011).

Reflections

In the institutional context of the school hierarchy, with teachers as decision-makers, pupils can choose whether to comply, or behave subversively, but have no real power to change the structures within which they are behaving (Barker et al., 2010). The children's researcher, in a school as a visitor, and on the goodwill of the school's management committee, finds themselves in a none-too-dissimilar position. An academic researcher occupies an uneasy position within the school hierarchy, particularly when they are in the school at their own behest, and conducting research that is negotiated directly with the school, and not explicitly supported by the Local Education Authority. Whilst researchers may be viewed by the children as 'responsible adults' and similarly expected to behave in a responsible manner by school staff, the researcher has little control over what occurs in the school environment, outside of the immediate research relationship or activity they are producing. The vignettes presented in this paper suggest that even then, control over their own research agendas may be fragile.

The researcher and the school staff are likely to have different desired outcomes from their interactions with the young people. The distinction between viewing the children as 'pupils' or as 'participants' can have significant implications for methods and ethics. However, the breadth of the gulf in perception of the children is likely to vary enormously dependent on the ethos and approach of the school and the researcher. Whilst school staff's primary motivation is to provide educational opportunities, typically child researchers are more concerned with the rights of the child participant (e.g. Cope, 2008). In an institutional context, actions that best support the goals of the institution will take priority over those that promote the ethos of the research. As such, maintaining an approach that promoted the agency of the participants proved impossible in some circumstances in both the school context, and also in research in the home.

In retrospect, the difference in the approaches, and the resulting unpleasant experience in the classroom for Ryan, might have been avoided by a more honest discussion of my approach to research with gatekeepers. In the particular case of Ryan being 'forced' to participate, I think the class teacher took my acquiescence to Ryan's request to miss the session as being too 'soft' or too laissez-faire. A more direct conversation at the outset, in which I explained the importance to the research of ongoing informed consent, and the reason I intended to conduct research in this manner, might have avoided the situation. Differences arising later on might have been discussed in the light of this early conversation. Similarly, in the situation of Dan, who was removed from the school camp, whilst I tried to advocate for him at the time, I didn't discuss the incident again later on. My reticence to have these, potentially difficult, conversations related to my ongoing awareness of being a 'guest' at the school's pleasure or in parents' homes (following Davis et al., 2000; Nilsen & Rogers, 2005). In relation to the school, I had approached over 20 institutions that I could commute to from my home, and Broadheath High was the only school that had responded positively to use the school as a base for research. Similarly, in terms of the research in family homes, I had sent information to around 80 families, of which only 12 had positively responded. As such, I suppose I felt an ongoing debt to the school, and the participant families and a wariness of the fragility of the whole of my PhD project resting on a successful outcome of research within this one particular school, and these few families.

Whilst the differing approach to the children as 'participants' and 'sons or daughters' might not be as far removed as the participant/pupil binary, there are likely to be some conflicts in the approach between researcher and parents. For example, households usually have rules that place boundaries on acceptable behaviour. These rules, whether stipulated or unspoken, may involve the ways in which it is appropriate to use your body, interact with others or dictate expected family practices such as mealtimes (e.g. Curtis, James, & Ellis, 2010). They are likely to involve some degree of awareness of a hierarchy of power in which children do as they are told by their parents, and by extension are expected to be polite and mindful of other adults. This was well illustrated through James' mother's decision-making regarding our trip, and her horror when she heard how James had challenged my authority during the visit. As in the school, these patterns of behaviours, or cultural norms of the household, may conflict with the desired approach of the researcher who may seek to unsettle the ingrained power hierarchy and promote the children's own agency allowing children to refuse to participate or communicate in their preferred manner.

A further complication, as discussed in the story of Dan's expulsion from camp, was my constantly shifting position as a researcher, as I temporarily accepted alternative responsibilities, in part as a result of the liminal nature of my position. Whilst on camp, I had agreed to undertake the role of a member of staff and students were therefore expecting to follow orders I gave. In the moment that I watched Dan messing around, and reached for the camera, I was acting as researcher, keen to get some fun footage to add to the participants' video diaries and use for discussion later. The changing of my role was, I believe, also a contributing factor to the unpleasant situation with James. Working one-to-one with James in a safe and private environment, we'd have been able to focus on activities that were fun. He could be silly and relaxed, but was polite and communicative. I think the fact that I began to take an increasingly authoritarian, rather than fun and friendly, role on our daytrip, as James became increasingly excitable was a surprise and a concern to him.

Whilst the incidents with Dan and Ryan highlighted the vulnerability of a researcher working within the institutional bounds of the school, James' story illustrates the vulnerability inherent from the increased responsibility associated with research that occurs outside institutional boundaries. On the one hand, working directly with families provided greater freedom and control over the research process. However, it also required a high degree of responsibility for the welfare of the participants. Part of this responsibility was the need to directly address unacceptable behaviour, primarily for practical and safety reasons (cf. Punch, 2007). Within the context of the school, unruly behaviour was not my responsibility to deal with. The contrasting isolation in the research trip also led to the potential for a high degree of vulnerability when I was accused by James of assault. I was lucky that James' parents didn't believe him. However, I was also lucky that I had a second adult present. Had James' parents believed his story, Lucy might have proved an invaluable witness.

The analysis of these events places a heavy emphasis on unidirectional power relations, which risk reinforcing the constructions of adult–child relations that lead to these concerning incidents in my research. To a large degree, this is intentional; the incidents arose in situations in which adults were exerting control over the young people. Further, my research with learning disabled children in school, and in their homes, found a high level of control over their day-to-day lives, suggested by other authors to be far more pervasive than for non-disabled children (Morris, 1997). However, it is important to also recognise the incidences of subversion in which the children used their own agency to rebel against adult control, demonstrating that learning disabled children may not always conform to their expected roles as compliant in the acceptance of adult control over their bodies, minds and behaviours (Davis et al., 2000). Indeed, the three incidents presented in the vignettes, all became moments of discomfort because the young people chose to exercise their own agency, whether through ignoring instructions, requesting to withdraw consent or 'messing around'.

The difference place makes

The three vignettes describe research experiences in schools, at home and in 'alternative' locations: on camp and at a nature reserve. The stories serve to highlight the importance of the institutional context of the research, and the fact that the role of the institution can complicate the microgeography of the research encounter. For example, the small group session that Ryan participated in took place in a separate classroom where I was allowed to conduct sessions as I chose. However, ultimate control remained in the hands of the classroom teacher who oversaw the research and decided who should be given the opportunity to take part, meanwhile limiting participants' rights to opt out. Similarly, Dan's actions, shouting and messing about in the pool, were not inappropriate in the space in which they occurred, an outdoor pool during leisure time. However, they were labelled as inappropriate as a result of the institutional context surrounding the event. Just as Maillet et al. (2016) describe state borders as geographical boundaries that move with asylum seekers, ensuring that they are always kept outside the territorial zone, institution boundaries also moved with the staff and pupils and with a research relationship established through school. In this way, the residential camp still might be argued to have occurred within the boundaries of school, whilst being geographically situated many miles from the physical school site. Despite the microgeography of the research encounter, in these cases, the institutional context was the overall determinant of the research relationship.

This paper has sought to provide an honest reflection on the experiences of conducting research within schools and families, particularly focusing on the importance of the site in which the research occurs. The use of three vignettes provides an embarrassing window into some of my most difficult experiences as a researcher. However, the subsequent reflection seeks to put these incidents into context of the conflict between the motivations of the researcher and school staff and parents, in relationships and specific interactions between adults and children. Whilst I don't claim to have the 'solution' to the potential difficulties of these relationships, I would advocate for clear, direct and honest communication between the researcher and the children's 'gatekeepers' over the intentions of the research relationship, even if this risks the viability of the research placement.

Acknowledgements

I thank the young people and teaching staff who participated in this research. Also thanks to my supervisors, James Evans, Fiona Smyth and Mark Jayne. The paper was much improved through the kind and thoughtful comments of the anonymous peer reviewers.

Disclosure statement

No potential conflict of interest was reported by the author.

Funding

This work was supported by the Economic and Social Research Council [grant number ES/G014906/1].

References

Abebe, T., & Bessell, S. (2014). Advancing ethical research with children: Critical reflections on ethical guidelines. *Children's Geographies, 12*, 126–133.

Alderson, P. (1995). *Listening to children: Children, ethics and social research*. Ilford: Barnardos.

Allen, L. (2009). 'Caught in the act': Ethics committee review and researching the sexual culture of schools. *Qualitative Research, 9*, 395–410.

Anderson, J., & Jones, K. (2009). The difference that place makes to methodology: 'Uncovering the 'lived space' of young people's spatial practices. *Children's Geographies, 7*, 291–303.

Barker, J., Alldred, P., Watts, M., & Dodman, H. (2010). Pupils or prisoners? Institutional geographies and internal exclusion in UK secondary schools. *Area, 42*, 378–386.

Bell, S., Phoenix, C., Lovell, R., & Wheeler, B. W. (2015). Using GPS and geo-narratives: A methodological approach for understanding and situating everyday green space encounters. *Area, 47*, 88–96.

Booth, K. (2015). What a difference place makes: Place gestalt and some methodological thoughts. *Qualitative Inquiry, 21*, 20–27.

Bushin, N. (2007). Interviewing the children in their homes: Putting ethical principles into practice and developing flexible techniques. *Children's Geographies, 5*, 235–251.

Buttimer, J., & Tierney, E. (2005). Patterns of leisure participation among adults with mild intellectual disability. *Journal of Intellectual Disabilities, 9*, 25–42.

Collins, D., & Kearns, R. (2001). The safe journeys of an enterprising school: Negotiating landscapes of opportunity and risk. *Health and Place, 7*, 293–306.

Contact A Family. (n.d.). *Research*. Retrieved October 20, 2016, from http://www.cafamily.org.uk/get-involved/campaigns-and-research/research/

Cooney, G., Jahoda, A., Gumley, A., & Knott, F. (2006). Young people with learning disabilities attending mainstream and segregated schooling perceived stigma, social comparisons and future aspirations. *Journal of Intellectual Disability Research, 50*, 432–444.

Cope, M. (2008). Becoming a scholar-advocate: Participatory research with children. *Antipode, 40*, 428–435.

Costley, D. (2000). Collecting the views of young people with moderate learning difficulties. In A. Lewis & G. Lindsay (Eds.), *Researching children's perspectives* (pp. 163–172). Buckingham: Oxford University Press.

Curtis, P., James, A., & Ellis, K. (2010). Children's snacking, children's food: Food moralities and family life. *Children's Geographies, 8*, 291–302.

Davis, J., Watson, N., & Cunningham-Burley, S. (2000). Learning the lives of disabled children. In P. Christensen & N. James (Eds.), *Research with Children* (pp. 201–224). London: Falmer.

Denscombe, M., & Aubrook, L. (1992). 'It's just another piece of school work': The ethics of questionnaire research on pupils in schools. *British Educational Research Journal, 18*, 113–131.

Dorrer, N., McIntosh, I., Punch, S., & Edmond, R. (2010). Children and food practices in residential care: Ambivalence in the 'institutional' home. *Children's Geographies, 8*, 247–259.

Dyer, S., & Demeritt, D. (2009). Un-ethical review? Why it is wrong to apply the medical model of research governance to human geography. *Progress in Human Geography, 33*, 46–64.

Elwood, S., & Martin, D. (2000). "Placing" interviews: Location and scales of power in qualitative research. *Professional Geographer, 52*, 649–657.

Evans, J., Evans, B., & Rich, E. (2003). 'The only problem is, children will like their chips': Education and the discursive production of ill-health. *Pedagogy, Culture and Society, 11*, 215–240.

Goodfellow, A. (2012). Looking through the learning disability lens: Inclusive education and the learning disability embodiment. *Children's Geographies, 10*, 67–81.

Griffin, K., Lahman, M., & Opitz, M. (2014). Shoulder-to-shoulder research *with* children: Methodological and ethical considerations. *Journal of Early Childhood Research*, *14*, 18–27.

Halse, C., & Honey, A. (2007). Rethinking ethics review as institutional discourse. *Qualitative Inquiry, 13*, 225–352.

Hammersley, M. (2009). Against the ethicists. *International Journal of Social Research Methodology, 12*, 211–225.

Hemming, P. (2011). Meaning encounters? Religion and social cohesion in an English primary school. *Social and Cultural Geography, 12*, 63–81.

Holloway, S., & Pimlott-Wilson, H. (2014). Enriching children, institutionalizing childhood? Geographies of play, extracurricular activities and parenting in England. *Annals of the Association of American Geographers, 104*, 613–627.

Harden, J. (2012). Good sitting, looking and listening: The regulation of young children's emotions in the classroom. *Children's Geographies, 10*, 83–93.

Holt, L. (2003). (Dis)abling children in primary school micro-spaces: Geographies of inclusion and exclusion. *Health and Place, 9*, 119–128.

Holt, L. (2004). Children with mind-body differences: Performing disability in primary school classrooms. *Children's Geographies, 2*, 219–236.

Holton, M., & Riley, M. (2014). Talking on the move: Place-based interviewing with undergraduate students. *Area, 46*, 59–65.

Hood, S., Kelley, P., & Mayall, B. (1996). Children as research subjects: A risky enterprise. *Children and Society, 10*, 117–128.

Horton, J. (2001). Do you get some funny looks when you tell people what you do?' Muddling through some angsts and ethics of (being a male) researching with children. *Ethics, Place & Environment, 4*, 159–166.

Horton, J. (2008). A 'sense of failure'? Everydayness and research ethics *Children's Geographies, 6*, 363–383.

Jahonda, A., Wilson, A., & Stalker, K. (2010). Living with stigma and the self-perceptions of people with mild intellectual disabilities. *Journal of Social Issues, 66*, 521–534.

Jones, K. (2008). 'It's well good sitting in the storecupboard just talking about what we do': Considering the spaces/places of research within children's geographies. *Children's Geographies, 6*, 327–332.

Klocker, N. (2012). Conducting sensitive research in the present and past tense: Recounting the stories of current and former child domestic workers. *Geoforum, 43*, 894–904.

Lewis, R. (2009). Recruiting parents and children into a research project: A qualitative exploration of families' decision-making processes. *International Journal of Social Research Methodology, 12*, 405–419.

Malone, K. (2007). The bubble-wrap generation: Children growing up in walled gardens. *Environmental Education Research, 13*, 513–527.

Maillet, P., Mountz, A., & Williams, K. (2016). Researching migration and enforcement in obscured places: Practical, ethical and methodological challenges to fieldwork. *Social and Cultural Geography*. doi: 10.1080/14649365.2016.1197963

Mandell, N. (1988). The least-adult role in studying children. *Journal of Contemporary Ethnography, 16*, 433–467.

Masson, J. (2000). Researching children's perspectives: Legal issues. In A. Lewis & G. Lindsay (Eds.), *Researching children's perspectives* (pp. 34–45). Buckingham: Oxford University Press.

Morris, J. (1997). Gone missing? Disabled children living away from their families. *Disability and Society, 12*, 241–258.

Matthews, H., Limb, M., & Taylor, M. (2000). The 'street as thirdspace'. In S. Holloway & G. Valentine (Eds.), *Children's geographies: Living, playing, learning and transforming everyday worlds* (pp. 54–68). London: Routledge.

Mauthner, M. (1997). Methodological aspects of collecting data from children: Lessons from three research projects. *Children and Society, 11*, 16–28.

McMaugh, A. (2011). En/countering disablement in school life in Australia: Children talk about peer relations and living with illness and disability. *Disability and Society, 26*, 853–866.

Mills, S. (2013). Surprise! Public historical geographies, user engagement and voluntarism. *Area, 45*, 16–22.

Morris-Roberts, K. (2001). Intervening in friendship exclusion? The politics of doing feminist research with teenage girls. *Ethics, Place and Environment, 4*, 147–153.

Morris-Roberts, K. (2004). Girls' friendships, 'distinctive individuality' and socio-spatial practices of (dis) identification. *Children's Geographies, 2*, 237–255.

Morrow, V., & Richards, M. (1996). The ethics of social research with children: An overview. *Children and Society, 10*, 90–105.

Nilsen, R., & Rogers, B. (2005). 'That's not a good idea, mom': Negotiating children's subjectivity while constructing 'home' as research site. *Children's Geographies, 3*, 345–362.

Oliver, M. (2004). If I had a hammer: The social model in action. In J. Swain, S. French, C. Barnes, & C. Thomas (Eds.), *Disabling barriers – Enabling environments* (pp. 7–12). London: Sage.

Pike, J. (2010). 'I don't have to listen to you! You're just a dinner lady!': Power and resistance at lunchtimes in primary schools. *Children's Geographies, 8*, 275–287.

Punch, S. (2007). 'I felt they were ganging up on me': Interviewing siblings at home. *Children's Geographies, 5*, 219–234.

Pyer, M. (2008). Unintended consequences? Exploring the un(fore)seen effects and outcomes of research. *Children's Geographies, 6*, 213–217.

Pyer, M., & Tucker, F. (2014). 'With us, we, like, physically can't': Transport, mobility and the leisure experiences of teenage wheelchair users. *Mobilities*, 1–17. In press

Robson, E., & Ansell, N. (2000). Young carers in Southern Africa: Exploring stories from Zimbabwean secondary school students. In S. Holloway & G. Valentine (Eds.), *Childrens Geographies* (pp. 174–93). London and New York, NY: Routledge.

Robinson, C. (2009). Nightscapes and leisure spaces: An ethnographic study of young people's use of free space. *Journal of Youth Studies, 12*, 501–514.

Rose, G. (1997). Situating knowledges: Positionality, reflexivities and other tactics. *Progress in Human Geography, 21*, 302–320.

Shahtahmasebi, S., Berridge, D., Lancaster, G., & Emerson, E. (2011). Child disability and the dynamics of family poverty, hardship and financial strain: Evidence from the UK. *Journal of Social Policy, 40*, 653–673.

Solish, A., Perry, A., & Minnes, P. (2010). Participation of children with and without disabilities in social, recreational and leisure activities. *Journal of Applied Research in Intellectual Disabilities, 23*, 226–236.

Starkweather, S. (2012). Telling family stories: Collaborative storytelling, taking precedence and giving precedence in family group interviews with Americans in Singapore. *Area, 44*, 289–295.

Tucker, F., & Matthews, H. (2001). 'They don't like girls hanging around there': Conflicts over recreational space in rural Northamptonshire. *Area, 33*, 161–168.

Tyrrell, N., & Harmer, N. (2015). A good move? Young people's comparisons of rural and urban living in Britain. *Childhood, 22*, 551–565.

Ursin, M. (2011). 'Wherever I lay my head is home' – Young people's experience of home in the Brazilian street environment. *Children's Geographies, 9*, 221–234.

Valentine, G. (1999). Being seen and heard? The ethical complexities of working with children and young adults at home and school. *Philosophy and Geography, 2*, 141–155.

Valentine, G., & Skelton, T. (2003). Living on the edge: The marginalisation and 'resistance' of D/deaf youth. *Environment and Planning A, 35*, 301–321.

van Blerk, L. (2005). Negotiating spatial identities: Mobile perspectives on street life in Uganda. *Children's Geographies, 3*, 5–21.

von Benzon, N. (2011). Who's afraid of the big bad woods? Fear and learning disabled children's access to nature. *Local Environment, 16*, 1021–1040.

Winchester, H. (1996). Ethical issues in interviewing as a research method in human geography. *Australian Geographer, 2*, 117–131.

Winter, K. (2012). Ascertaining the perspectives of young children in care: Case studies in the use of reality boxes. *Children and Society, 26*, 368–380.

Wolpert, J. (1980). The dignity of risk. *Transactions of the Institute of British Geography, 5*, 391–401.

Worth, N. (2013). Making friends and fitting in: A social-relational understanding of disability at school. *Social and Cultural Geography, 14*, 103–123.

Index